SpringerBriefs in Electrical and Computer Engineering

Series editors
Woon-Seng Gan
Sch of Electrical & Electronic Engg
Nanyang Technological University
Singapore, Singapore

C.-C. Jay Kuo
University of Southern California
Los Angeles, California, USA

Thomas Fang Zheng
Res Inst Info Tech
Tsinghua University
Beijing, China

Mauro Barni
Dept of Info Engg & Mathematics
University of Siena
Siena, Italy

More information about this series at http://www.springer.com/series/10059

Xavier Fernando • Ajmery Sultana • Sattar Hussain
Lian Zhao

Cooperative Spectrum Sensing and Resource Allocation Strategies in Cognitive Radio Networks

 Springer

Xavier Fernando
Department of Electrical and Computer
Engineering
Ryerson University
Toronto, ON, Canada

Ajmery Sultana
Department of Electrical and Computer
Engineering
Ryerson University
Toronto, ON, Canada

Sattar Hussain
Department of Information and
Communication Engineering Technology
Centennial College
Toronto, ON, Canada

Lian Zhao
Department of Electrical and Computer
Engineering
Ryerson University
Toronto, ON, Canada

ISSN 2191-8112   ISSN 2191-8120   (electronic)
SpringerBriefs in Electrical and Computer Engineering
ISBN 978-3-319-73956-4   ISBN 978-3-319-73957-1   (eBook)
https://doi.org/10.1007/978-3-319-73957-1

Library of Congress Control Number: 2018939291

Printed on acid-free paper

This Springer imprint is published by the registered company Springer International Publishing AG part
of Springer Nature.
The registered company address is: Gewerbestrasse 11, 6330 Cham, Switzerland

Preface

The use of cognition in radio access and networking is expected to enable a number of significant enhancements in wireless communications. In addition to better utilization of spectrum, other not so obvious areas such as autonomous network configuration, interference reduction, energy efficiency, interoperability, and coexistence among different wireless/mobile communications systems and devices can also benefit from a cognitive approach. The cognitive and adaptive capabilities of radio access will be of fundamental importance in 5G and beyond networking contexts especially with heterogeneous networking and massive multiple input, multiple output (MIMO) scenarios envisioned in 5G. Cognitive networking will enable future networks to become more adaptive, self-configuring, self-organizing, self-healing, and self-recovering. It will enable networks to cope with environmental changes, network dynamics, and malicious attacks, etc.

As a first step, cognitive radio (CR) technology in the radio access networks is widely researched that could solve the spectrum scarcity issue. Ever-increasing demand for enhanced wireless services with wire line quality postulates higher data rates, even up to tens of Gbps and pushes the spectrum to its limits. While there is constant push for new spectrum, current licensed spectrum is significantly underutilized. CR approach will provide efficient spectral usage using intelligent wireless nodes. For the successful implementation of CR networks, accurate spectrum sensing in real time, fast adaptation, and efficient resource allocation (RA) schemes are essential. Due to the uncertainty of the varying multipath wireless channels, often cooperative spectrum sensing using multiple secondary nodes is used for a more reliable solution.

In this book, the cognitive approach in radio access networks is introduced first. Accurate spectrum sensing is essential for the successful implementation of CR networks (CRN). Then, a framework for studying the performance of real-time *cooperative* spectrum sensing methods is described in CRN by considering realistic channel conditions. The analysis aims to show that the effects of multipath fading can be alleviated by using relay-based cooperative spectrum sensing approaches. The sensing time, end-to-end SNR, and end-to-end channel statistics are considered into the performance analysis of cooperative sensing CRNs. An advanced statistical approach is introduced to derive new exact closed-form expressions for average false alarm probability and average detection probability in this scenario. A novel approximation to alleviate the computational complexity of the proposed models is also discussed.

Once the spectral holes are identified with accurate spectrum sensing, efficient and systematic RA shall be performed for optimal resource usage. In the second part of the book, the taxonomy for the RA process in CRNs is provided. The RA problems are categorized first according to the adopted approaches and network architecture. Then, the optimization strategies for the RA are explored and reviewed in the context of CRNs. Finally, a comprehensive overview of these categories and optimization strategies is provided, and the advantages, disadvantages, and the application areas associated with each optimization strategy are highlighted. The device-to-device (D2D) scenario is discussed as a case study and the application of the CR technology in the D2D realm in more detail. Various optimization strategies are reviewed to solve the RA problems in the context of CRNs. The application of advanced geometric water-filling (GWF) approach with more stringent constraints is presented in detail in this book. Numerical results are also provided to reveal more insights quantitatively.

Toronto, ON, Canada   Xavier Fernando
October 2017   Ajmery Sultana
                                                                              Sattar Hussain
                                                                                  Lian Zhao

Contents

Abbreviations

AF	Amplify-and-forward
AWGN	Additive white gaussian noise
CCI	Co-channel interference (CCI)
CDF	Cumulative distribution function
CF	Characteristic function
CR	Cognitive radio
CRC	Cyclic redundancy check
CRN	Cognitive radio network
CRP	CR pair
CSI	Channel state information
CSS	Cooperative spectrum sensing
CWF	Conventional water-filling
CXI	Cross-channel interference (XCI)
DF	Decode-and-forward
DFT	Discrete Fourier transform
DSL	Digital subscriber lines
DVB	Digital video broadcasting
D2D	Device-to-device
EGC	Equal gain combining
FEC	Forward error-correction
FFT	Fast Fourier transform
GWF	Geometric water-filling
GWFPP	Geometric water-filling with individual peak power constraints
IGPP	Iterative partitioned weighted geometric water-filling with individual peak power constraints
i.i.d	Independent identically distributed
IPWF	Iterative partitioned water-filling
ISI	Inter-symbol interference
LOS	Line of sight
LTE	Long term evaluation
LWF	Linear water-filling

MAC	Medium access control
MGF	Moment generating function
MIMO	Multiple input, multiple output
MRC	Maximal ratio combining
OFDM	Orthogonal frequency division multiplexing
PDF	Probability density function
PHY	Physical layer
PU	Primary user
PUP	PU pair
QoS	Quality of service
RF	Radio frequency
ROC	Receiver operating characteristic
RRA	Radio resource allocation
SC	Selection combining
SDR	Software-defined radio
SNR	Signal-to-noise ratio
SU	Secondary user
TDMA	Time division multiple access
UWB	Ultra wide band
WRAN	Wireless regional area network

Chapter 1
Introduction

Cognitive radio technology is emerging as an attractive solution to spectral conges-
tion problem by enabling unlicensed (or secondary) users to coexist with licensed
(or primary) users without harmful interference to the primary users who is entitled
to use the frequency band.

In fact, the coexistence of primary and secondary users (PUs and SUs) in a CR
environment are generally categorized into three paradigms [1]:

Underlay: The SU determines the interference caused by their transmission to
    the PU and transmits only if interference below (under) a given threshold. The
    interference constraint may be met by using wideband techniques such as spread
    spectrum or Ultra Wide Band (UWB) or by directional radiation.
Interweave: Interweave systems completely avoid interference by not transmit-
    ting in a band that is occupied by the PU.
Overlay: In overlay systems, simultaneous transmission is allowed. Here, the
    cognitive user has the knowledge of PU's message and/or encoding strategy.
    Simultaneous communication is achieved via various encoding and interference
    mitigation approached.

However, in general CR systems are assumed to be interweave. In this book,
we exclusively consider interweave systems that opportunistically use of frequency
bands that are not completely occupied by the licensed users. In interweave systems,
the SUs shall periodically monitor the spectrum for vacant channels to start their
own communications and vacate the spectrum if the PU starts transmission. A
fundamental requirement is that CR users should constantly and accurately sense
the spectrum occupancy.

Since, a single SU might miss detect (due to shadowing and hidden node issues),
cooperative spectrum sensing is often deployed where the sensing information from
multiple SUs is gathered at a fusion center (gaining from sensing diversity) to make
an accurate estimate.

© The Author(s), under exclusive licence to Springer International Publishing AG,   1
part of Springer Nature 2019
X. Fernando et al., *Cooperative Spectrum Sensing and Resource Allocation
Strategies in Cognitive Radio Networks*, SpringerBriefs in Electrical and Computer
Engineering, https://doi.org/10.1007/978-3-319-73957-1_1

This chapter presents an overview of interweave cognitive radio networks (CRNs) and the advantages of implementing cooperative spectrum sensing (CSS) in these networks. The chapter discusses spectrum sensing methods and gives an overview of diversity combining techniques that can be used to improve detection accuracy. It also investigates challenges in the implementation of CSS in both centralized and distributed cooperative sensing networks.

1.1   The Interweave Cognitive Radio System

Ever increasing demand for myriad of wireless services poses two major challenges in the wireless network paradigm. One is the spectrum scarcity and the other is the demand of high data rates, up to tens of Gbps. While there is continuous effort to allocate more spectrum for wireless usage, it is observed that currently licensed spectrum is significantly underutilized [2, 3] due to sporadic transmission nature of most communication devices.

CR concept was first coined by Joseph Mitola [4], who proposed it as a solution for efficiently utilizing the radio resources. Original CR concept was strictly *interweave.* Since a CR transceiver shall have the ability to tune to different frequency bands, the CR is typically built using software-defined radio technology. Therefore, the transmitter operating parameters, such as the carrier frequency, modulation type and transmission power can be dynamically adjusted by software [5, 6]. CRs, with its ability to smartly interact with the surrounding environment, are amenable to allow the coexistence licensed users (PUs) and unlicensed users (SUs) sharing the same bandwidth opportunistically without causing harmful interference to each other.

Spectrum sensing is one of the most fundamental elements of a CRN; the task can be seen as occupying two-layers. The PHY-layer detection methods such as energy detection, matched filter, and feature detection aim to efficiently discover the presence of primary user signal by learning its modulation/encoding schemes and parameters. On the other hand, the MAC-layer sensing determines how frequently and when, the secondary user has to sense given channels (sensing period). The sensing period has to be optimized based on the traffic characteristics of the PUs as well as SUs [7].

Spectral hole locations usually change dynamically. This is an issue in optimizing the sensing period and need wideband agile receivers with fast sensing capabilities. Advanced techniques such as compressed sensing may be helpful here. Also note, sensing operation drains energy which can be an issue if the secondary network is a wireless sensor network.

CRNs have distinctive characteristics from a traditional wireless network where it intelligently recognizes the status of the radio environment and adjusts its functional parameters accordingly [6, 8]. Most critical part of CRN is allowing CR users to share the licensed spectrum with PUs without degrading their performance [7]. This imposes new challenges and open research issues.

Fig. 1.1 Basic elements of
cognitive radio operation

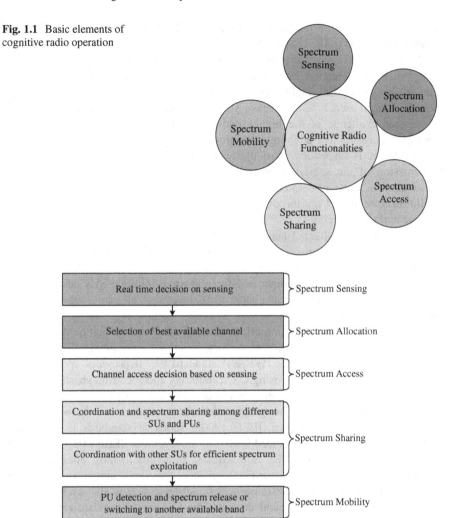

Fig. 1.2 Cognitive radio functionalities

The basic elements for the operation of CR systems are depicted in Fig. 1.1 and
the main functions of those elements to support intelligent and efficient utilization
of frequency spectrum are envisioned in Fig. 1.2.

From Figs. 1.1 and 1.2, it can be seen that the proper functionality of a CRN
depends on optimally sensing, accessing and sharing operations of the spectrum.
Capacity of CR depends on what white spaces are available and how accurately
they are detected. In addition to fading and shadowing, random interference can also
cause false detection. Furthermore, SU transmitter and receiver must coordinate to
find common holes for successful transmission. Therefore, the cognitive network
throughput is usually difficult to guarantee. Ergodic and outage capacity of a CR

system depends on the percentage of successfully detected white spaces which depends on the PU traffic characteristics.

Usually, the SU shall keep sensing the spectrum even when they transmit. However, implementation of this in a satisfactory manner is not easy as most transceivers are not full duplex. This can cause unintended interference to the PU. Hence, licensed users hate interweave SUs even more than underlay SUs.

Another challenge is the higher end-to-end delay in CRNs caused by the sensing, processing and adapting operations. During the spectrum sensing period payload traffic is suspended. Particularly, when cooperative sensing is performed, transmission of the sensed information to the coordinator and processing this information to make an optimal decision consumes time processing capacity. This would directly impact the throughput, quality of service (QoS) and increases the average end-to-end delay.

Due to the multi-channel nature of CRN, the multi-dimensional nature of the RA optimization problems become very difficult to solve. In addition to finding the optimal subcarrier set and power allocation for each SU, another challenge would be to avoid the co-channel interference when multiple SUs decide to use the same frequency band. Hence, a better MAC layer shall also cater the co-channel interference among the SUs [1].

Regulation and standardization efforts have already been carried out to envision some of the applications of CRN. A very important example is IEEE 802.22 wireless regional area network (WRAN) standard [9]. It provides specifications for broadband wireless access using CR technology and spectrum sharing policies and procedures for operation in the white space TV bands. IEEE 802.11af standard and its amendments [10] enable geo-location database access in the white space radio frequency (RF) spectrum. IEEE 1900.x series of standards [11] provide next generation radio and advanced spectrum management. IEEE 802.19 standard [12] enables the family of IEEE 802 wireless standards to most effectively use TV white space by providing standard coexistence methods among dissimilar or independently operated IEEE 802 networks and devices. It is also useful for non IEEE 802 networks and TV band devises. The first set of standardization study towards licensed shared access for long term evaluation (LTE) is reported in [13] that were successfully tested in a live LTE network in the 2.4–2.5 MHz frequency band [14]. In most recent times, IEEE has engaged the 802.15.4m task group [15] to characterize cognitive radio-aware PHY and MAC layers for cognitive machine-to-machine networks. These examples show standardization process for CR is well underway.

Latest developments in spectrum allocation policy and regulatory domains, including the release of the National Broadband plan, the publication of final rules for TV-white spaces, and the ongoing proceeding for secondary use of the 2360–2400 MHz band for medical body area networks, have opened up various opportunities for the secondary use of spectrum. CR is therefore addressed to enable and support a variety of emerging applications, ranging from smart grid, public safety and broadband cellular, medical applications to WSNs. The wide range of CR applications have various design requirements. For instance, from the data rate point of view, a temperature sensor may have a very low data rate whereas cellular handset may have a very high data rate. The IEEE 802.15.4 standard supports different data rates varying from 20 kbps (868 MHz band) to 250 kbps (2.4 GHz band).

1.2 Cooperative Spectrum Sensing in CRNs

Reliable detection of the existence of PUs is a primary requirement for minimization of interference to the primary network [16]. In a real communication environment, the local sensing performance of individual users may severely degrade due to deep fading/shadowing and hidden node issues [17–19]. Therefore, individual spectrum sensing is unreliable and prone to errors. Generally, reliable spectrum sensing is a critical and challenging issue due to the nature of the wireless channels in fading and shadowing environments [20, 21]. Fortunately, the accuracy of detecting the primary network can be improved by sharing the local observations among CR users which adds diversity to spectrum sensing networks and helps improve the detection reliability. By exploiting the diversity provided by associated radios, CSS improves the overall detection sensitivity without imposing higher sensitivity requirements on the individual CRs [17, 22, 23]. It has shown that a network of cooperative CRs [24, 25], which experience different channel conditions from the target, would have a better chance of detecting the primary radio if the individuals local sensing are jointly combined at a fusion center. CSS has also shown to reduce the detection time and increase the overall system agility [26, 27].

The basic idea behind cooperative transmission is that the signal transmitted by a source to a destination, each employing single antenna, is also received by other terminals which are often referred to as *relays* [28]. The destination then combines the signals coming from the source and the relays, thereby creating spatial diversity by taking advantage of the multiple receptions of the same data at the various terminals and transmissions paths. More specifically, the presence of multiple radios helps to reduce the effects of severed multipath at a single radio since they provide the destination with multiple independent realizations of related random variables. With multiple realizations, the probability that all users see deep fades is very low [29].

The merit of CSS primarily lies in the achievable space diversity brought by the sensing channels, namely, sensing diversity gain provided by the multiple CRs. Even though one CR may fail to detect the primary signal, there are still many chances for other CRs to detect it. Another merit of cooperative spectrum sensing is the mutual benefit brought forward by communicating with each other to improve sensing performance and system agility. When a CR is far away from a primary user, the received signal is too weak to sense its presence. But if another CR radio is located nearby the primary radio, then it can act as a relay to forward the signal of the primary user where it can be reliably detected by the far user in a very short time thereby improving system agility [26, 27]. Moreover, by allowing multiple CRs to cooperate in spectrum sensing, the hidden terminal problem can also be addressed [25]. While a cooperative approach provides a more accurate sensing performance, it causes adverse effects on resource-constrained networks due to the additional operations and overhead traffic [30, 31].

Obviously, CSS network communications go through two successive channels [28]: First, the sensing channels which are the links between the primary user and the CRs, and second, the relaying (reporting) channels which are the links between the CRs and the fusion center. For cooperative decisions, relaying channels are used by CR users to report their local observations (for soft-fusion decision) or their individual decisions (for hard-fusion decision) to a secondary base station (fusion center). In this dual-hop cooperative network, the primary user stands for the source and the fusion center stands for the destination. The basic structure of this cooperative approach is depicted in Fig. 1.3. In another approach of cooperative networks, secondary users with higher detection probabilities constantly act as relays to help those with lower detection probabilities [26, 27]. Such cooperation can reduce the detection time of the "weaker" user thereby improving the agility of the overall network. Figure 1.4 shows a scenario of two-user cooperation. In the presence of more than two users, radios can be grouped into pairs, in which one user acts as a relay for the other [27]. However, the performance of such fixed cooperation model will become worse when considering mobility of the CR users.

Fig. 1.3 Centralized cooperative spectrum sensing system

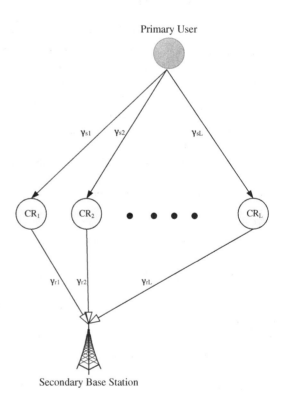

Fig. 1.4 Two-user
cooperation in cognitive radio

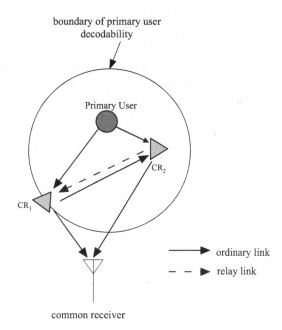

boundary of primary user
decodability

Primary User

CR₂

CR₁

ordinary link

- - ▶ relay link

common receiver

Relaying local sensing to a fusion center allows soft-fusion (data-fusion) policy to be used at the receiver [32]. The soft-fusion policy has been proved to achieves optimal diversity, providing a higher diversity gain and a better detection accuracy compared to the logic-based hard-fusion (also called decision-fusion) policy [32–34]. It also allows to keep the complexity of local sensors at a low level since only analog signals are required to be forwarded to the fusion center [33, 35]. However, the optimum performance comes on the cost of bandwidth requirements of the relaying links.

CSS can be implemented either in a centralized system or in a distributed system [36, 37]. In the centralized method, a base-station works as a fusion center to gather all sensing information from the CR users to detect the spectrum holes. Cooperation among CR users in a centralized manner is usually coordinated by a fusion center through either decision-fusion or soft-fusion policies [38, 39]. The centralized network architecture helps overcome the receiver uncertainty problem. With the transmitter detection, CR networks cannot avoid interference at the nearby primary receiver since the transmitter's detection relies only on local observations of the signal transmitted by a primary transmitter and may not aware of the location of the primary receiver [40]. But with the centralized system, the base-station can collect information about the locations of all primary users in its coverage area and provide such information to the CR users. The centralized approach also mitigates the limitation in sensing capabilities. For example, CR networks should be synchronized to schedule the spectrum sensing among all CR users. Thus, CR networks need to have a base-station to synchronize the scheduling such that all users can have the same sensing cycles and avoid interfere with sensing operations [41].

On the other side, distributed systems require an exchange of observations among CR users [42]. In the fully distributed system, all signal processing is done at the CR terminals. Decisions are taken locally by these terminals and forwarded later on to the fusion center. Global or final decision are then made by the fusion center [43]. Distributed network solutions are mainly proposed for cases where the construction of an infrastructure is not preferable [42, 44, 45]. These networks lack the centralized support and rely on local coordination for network management and topology configuration [46]. One common cooperative scheme of distributed systems is to form clusters to share the sensing information locally [47].

1.2.1 Challenges in Cooperative Spectrum Sensing Challenges

The basic assumption of the spectrum sensing is that the locations of the primary receivers are unknown due to the lack of signalling between primary users and CR users. CR terminals have to only rely on detecting weak primary transmitter signals based on their local observations [44]. Accordingly, CR receiver sensitivity must be higher than the highest primary receiver sensitivity by a large margin [30]. Even though, the detection of primary transmitter signals may not prevent a hidden terminal problem. The hidden terminal problem is considered as a critical challenging issue in the spectrum sensing [30]. It arises when the secondary user fails to detect the primary transmitter due to either deep fading/shadowing or receiver uncertainty as shown in Fig. 1.5. In this case, the CR cannot sense the

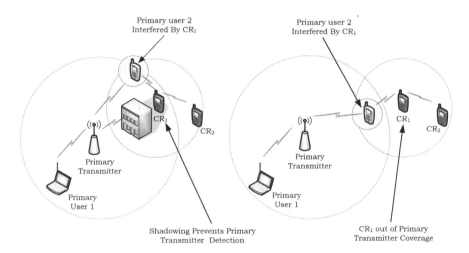

Fig. 1.5 Hidden terminal and receiver uncertainty problems

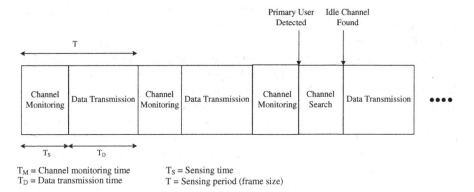

T_M = Channel monitoring time   T_S = Sensing time
T_D = Data transmission time   T = Sensing period (frame size)

Fig. 1.6 Periodic spectrum sensing structure

presence of the primary user signal, and thus it is allowed to access the channel while the primary user is still in operation.

Multipath fading is another main obstacle to achieve reliable communications by CR users [48]. Under channel fading or shadowing, a low received signal strength does not necessarily imply that the primary system is not active or located out of the CR's interference range as the signal may be experiencing a deep fade or is being heavily shadowed by obstacles [49]. Hence, a performance analysis for both coherent or noncoherent detections over fading channels requires knowledge of the fading envelope statistics and will be considered in detail in the following chapter.

CR users should monitor the spectrum continuously. Because the receiver's RF front-end cannot differentiate between the primary user signals and CR user signals, CR users can not transmit and sense at the same time [32, 50]. This necessitates a periodic sensing scheme where sensing and transmitting are alternating in a periodic manner with separate observation and transmission periods [41]. Figure 1.6, illustrates the basic structure of spectrum sensing cycles in CR networks. The periodic sensing structure introduces another design challenge [41]. Interference avoidance depends on the sensing accuracy which is in turn determined by the observation time. Since the transmission of CR users is not allowed during the observation time, the transmission time is reduced which inevitably reduces the spectrum usage efficiency.

Spectrum sensing is also challenged by the fact that the available spectrum holes show different characteristics which vary over time. In order to describe the dynamic nature of CR networks, each spectrum hole should be characterized by considering not only the time-varying radio environment but also the primary user activity and the spectrum band information such as operating frequency and bandwidth [30]. Most CR research assumes that the primary user activity is modeled by exponentially distributed inter-arrivals [41, 44, 51, 52]. In this model, the primary user traffic is modeled as a two state birth-death process. An ON (busy) state represents the period used by primary users and an OFF (idle) state represents the unused period [53]. Since each user's arrival is independent, each transition follows the Poisson arrival process and the length of the ON and the OFF periods are exponentially distributed.

Interference temperature introduces another challenge to CSS in CR networks. It has shown that interference temperature management technique may not lead to accurate detection [49, 54–56]. This technique accounts for the cumulative radio frequency energy from multiple transmissions, and sets limit on their aggregate level that the primary receiver could tolerate called the interference temperature limit (ITL) [57]. CR users are allowed to use the spectrum band as long as their transmission does not exceed the pre-defined limit. The distribution of the aggregate interference is characterized in terms of parameters such as sensitivity, transmitted power, density of cognitive radios, and the underlying propagation environment. The difficulty of this technique lies in accurately measuring the interference temperature since CR users cannot distinguish between actual signals from the primary user and noise/interference. Finally, the large operating bandwidths required by the spectrum sensing, impose additional requirements on the RF components such as antenna and power amplifiers as well as the need to high speed processing units (DSPs/FPGAs) for performing computationally demanding signal processing tasks with relatively low delay [58].

1.2.2 Spectrum Sensing Methods

Spectrum sensing techniques can be classified into three principal types [25]: matched filter detection, cyclostationary detection, and energy detection. When the noise is Gaussian and the signal has a known form, even with unknown parameters, the appropriate processing includes a matched filter or its correlator equivalent [59]. Matched filter is an optimal detector in AWGN channel since it maximizes the received SNR [60]. The main advantage of the matched filter is that it needs less time to achieve high processing gain due to coherent detection [61]. However, matched filters require a high synchronization between the primary user and the CRs [46]. Furthermore, CR users need to have different multiple matched filters dedicated to each type of the primary user signal, which increase the implementation cost and complexity [46].

Cyclostationary detection determines the presence of the primary user signals by extracting their specific features such as pilot signals, cyclic prefixes, symbol rate, spreading codes, or modulation types from its local observation [46]. The main advantage of the feature detection is its robustness to the uncertainty in noise power. If the signal of the primary user exhibits strong cyclostationary properties, it can be detected at very low SNR values by exploiting the information (cyclostationary features) embedded in the received signal [46, 62]. Furthermore this detection approach can distinguish the primary signal from other CR users' signals over the same frequency band provided that the cyclic features of the primary user and the CRs' signals differ from each other [63]. However, cyclostationary detection is more complex to implement than the energy detection and requires a prior knowledge of the primary signal such as modulation format [28, 64].

Energy detector is a non-coherent detector that is simple to implement [65]. The energy detector is a threshold device which measures the energy of the received waveform over an observation time window [65]. It is considered an optimal detector for unknown signal when the noise power is known [25]. Since the CR radio doesn't need to have prior knowledge about the primary signal, the energy detection is widely considered as a local spectrum sensing detection method. The reliability of the energy detector depends on the receiver's noise characteristics, the received signal strength, and the length of time that is used for the integration. While the energy detector is easy to implement and can be used without any prior knowledge of the primary signal, it still has some drawbacks. The first problem is that it has poor performance under low SNR conditions and can not be able to detect the signal reliably if the SNR is less than SNR_{wall} [25, 66]. SNR_{wall} is the weakest SNR below which a detector fails to reliably detect the signal. Since the threshold used in energy detection depends on the noise variance, any small noise power estimation errors can result in significant performance loss [32]. The energy detector also suffers from longer detection time compared to the matched filter detection [46]. Another challenging issue is the ability to differentiate the interference from other secondary users sharing the same channel and the primary user. For this reason, CR networks need to provide a synchronization over the sensing operations of all neighbors, i.e. each CR user should be synchronized with the same sensing and transmission schedules. Otherwise, CR users cannot distinguish the received signals from primary and CR users, and hence the sensing operations of the CR user will be interfered by the transmissions of its neighbors.

When the signal has an unknown form, it is sometimes appropriate to consider the signal as a sample function of a random process [59]. When the signal statistics are known, this knowledge can often be used to design a suitable detector. In the case of CR networks, it seems appropriate to use an energy detector to determine the activity of the primary network due to the absence of much knowledge concerning the primary user signals.

The energy detector consists of a square law device followed by a finite time integrator. First the input signal is filtered with a band-pass filter to select the bandwidth of interest. The filtered signal is then squared and integrated over a time interval, T. The requirement of T is that it must be short in comparison to the time required for the fading amplitude to change up appreciably, but long in comparison to the period of the signal [67]. Figure 1.7, depicts the block diagram of a typical energy detector.

For an arbitrary CR user, the signal receiver over the sensing link y_s at time t can be represented as

$$
\begin{aligned}
H_0: & \quad y_s(t) = n_s(t) \\
H_1: & \quad y_s(t) = h_s.x_p(t) + n_s(t)
\end{aligned}
\tag{1.1}
$$

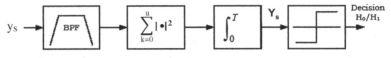

BPF = Band pass filter
T = Integrating time interval
u = Number of samples

Fig. 1.7 Typical energy detection receiver

where, x_p denotes the signal transmitted by the primary radio, h_s denotes the channel fading coefficient of the sensing link, and n_s denotes the additive Gaussian noise. The hypotheses H_0 and H_1 are for the primary signal's absence and presence, respectively. For local detection, the power of y_s denoted by Y_s is compared with a given threshold λ to infer the primary state following a decision rule given by [34]

$$Y_s \underset{H_0}{\overset{H_1}{\gtrless}} \lambda \qquad (1.2)$$

By definition, a false alarm occurs when the primary user activity is claimed under H_0, while an accurate detection occurs when the primary user activity is claimed under H_1. Therefore, the detection probability, P_d, and the false alarm probability, P_f, are given by,

$$P_d = P\{Y > \lambda | H_1\} \qquad (1.3)$$

$$P_f = P\{Y > \lambda | H_0\} \qquad (1.4)$$

Observation period or sensing time has a direct influence on sensing accuracy. For interference avoidance, the observation time needs to be long enough to achieve sufficient detection accuracy. But, increasing observation period inevitably decreases the transmission time and consequently decreases spectrum usage (sensing efficiency) since CR transmission is not allowed during observation period. On the other hand, a longer transmission time enhances the sensing efficiency but causes higher interferences to primary users as detection accuracy decreases due to the lack of sufficient sensing information [41]. Thus, a proper selection of transmission time and sensing time is crucial for the operation of energy detectors. Practically, there must be a trade off between interference avoidance and sensing efficiency.

1.2.3 Sensing and Relaying Channels

A relayed transmission is a well-known technique that has the advantage of extending the coverage without using large power at the transmitter. Recently, it has gained a new actuality in the context of collaborative wireless communication systems [27, 44, 68], where relaying is used as a form of spatial diversity to overcome highly shadowed or deeply faded links. As a result, the signal from the source to the destination propagates through two hops (links) in series [69].

If CR terminals communicate in a multi-hop fashion, they form a multi-hop cognitive radio network. The achievable detection accuracy of these networks depends on channel fading conditions, relaying mechanism, and combining scheme employed by the fusion center [70]. The difference between a conventional single-hop system and a relay-based dual-hop system is that the noise in the single-hop system is independent of the channel statistics while it is not the case in the dual-hop network. Importantly, the relaying channels convey noise to the destination, which is why it is necessary to incorporate its wireless propagation characteristics into the CRN's performance analysis.

The relaying channels are usually bandwidth limited [32]. If every user transmits the real value of its observation, a large relaying bandwidth is required which may not be available. The bandwidth limitations of the relaying channel can be addressed in two ways:

– Only the users with reliable information are allowed to report their local observations.
– Allow Users to report a binary decision to the common receiver instead of local observations.

Existing relaying techniques are classified into amplify-and-forward (AF) protocol and decode-and-forward (DF) protocol. In the AF protocol, CR users forward their raw measurements to the fusion center. While in the DF protocol, CR users perform local detections and only forward decisions to the fusion center. The latter case offers a substantial reduction in bandwidth requirements for reporting data to the fusion center. However, the AF protocol shifts the complexity from the local radios to the fusion center and offers a simple way to report raw data to the decision maker which has been shown to improve the spectrum sensing performance [28, 35]. Based on the AF gain, the relay is referred to as a CSI-assisted relay or as a fixed-gain relay. The former case requires the cognitive radio to be aware of the instantaneous sensing channel state to adjust its gain according to the sensed power. While the fixed-gain relay amplifies and forwards the received signal with a constant gain and consequently results in an output signal with a variable power. Although fixed-gain relaying significantly reduces the complexity of each relay, its performance in fading channels is still less explored than the CSI-assisted relay and is not expected to perform well as systems with CSI-assisted relays [71, 72]. In cognitive radio applications, the CSI may be available to secondary users over a control channel or over a broadcast channel through an access point. Several other relaying protocols have recently been investigated [73, 74].

Relaying links provide more flexibility for designing advanced cooperative protocols. The level of cooperation is determined by the bandwidth of the relaying channel. To this end, we can define two regimes of interest [49]:

– Low bandwidth relaying channels for energy detection and cyclostationary detection receivers: In this regime, it is realistic to assume that the CRs exchange decisions or summary statistics rather than long vectors of raw data.
– High bandwidth relaying channels: in this regime, CRs can exchange entire raw data and hence a sophisticated detection can be performed. In this scenario, it may be possible for tightly synchronized radios to collectively overcome poor detection scenarios.

When the relaying channel is jammed or unavailable, CR users should adapt to the new situation and use an alternative channel. For example, CR users may use a global control channel [75]. To facilitate the change over to the alternative relaying channel, efficient information sharing and coordination algorithms should be developed [76]. These algorithms must be designed to reduce overhead traffic caused by the cooperation process and should be robust to the temporal variations of relaying channels.

1.3   Diversity Combining Techniques

When the AF protocol is implemented, several diversity combining techniques can be used at the receiver. The objective of diversity combining techniques is to make use of the several received signals to improve the realized SNR and combat the effect of the multipath fading [67]. Diversity combining relies on a simple principle. Namely, if a number of well-separated relays are used to relay the same signal, it is unlikely that all received signals at the fusion center fade at the same time. In general, the term "diversity system" refers to a system in which one has available two or more closely similar copies of some desired signal [67]. Diversity combining offers one of the greatest potentials for radio link performance improvement [48].

Diversity schemes can be classified according to the type of combining employed at the receiver as MRC, SC, and EGC. The MRC combiner weighs the received signals with respect to their SNR and is known to be of high performance. However, MRC receiver's complexity is directly proportional to the number of diversity branches. The EGC equally weighs the diversity branches but it has the same undesirable feature of having the receiver complexity dependent on the number of diversity branches [77]. Conversely, the SC scheme selects only one relay which has the maximum channel gain among all the diversity branches. The combiner offers a simple receiver structure since it is simply performed by measuring the SNR of all the diversity branches.

When decision-fusion policy is used; AND, OR or M-out-of-N methods can be used for combining information from different cognitive radios [78]. However, in practical systems, to reliably deliver one-bit sensing information actually involves a lot of overheads, including reliable channel coding, packet preamble, CRC bits etc. Typically, the number of cooperative users and the amount of information that each user should send to the fusion center determine the required bandwidth. For a large network, the required bandwidth may exceed the capability of the reporting channels, which may be alleviated through DF relaying.

References

1. A. Sultana, X. Fernando, L. Zhao, An overview of medium access control strategies for opportunistic spectrum access in cognitive radio networks. Springer Peer-to-Peer Netw. Appl. **10**(5), 1113–1141 (2017)
2. Federal Communications Commission, Spectrum policy task force report, November 2002
3. G. Taylor, C. Middleton, X. Fernando, A question of scarcity: spectrum and Canada's urban core. J. Inform. Policy (2016). E-ISSN: 2158–3897
4. J. Mitola, Cognitive radio: an integrated agent architecture for software defined radio. Ph.D. thesis, KTH Royal Institute of Technology, Stockholm, 2000
5. I. Akyildiz, W. Lee, M. Vuran, S. Mohanty, Next generation/dynamic spectrum access/cognitive radio wireless networks: a survey. Elsevier Comput. Netw. **50**, 2127–2159 (2006)
6. S. Haykin, Cognitive radio: brain-empowered wireless communications. IEEE J. Sel. Areas Commun. **23**, 201–220 (2005)
7. A.S. Zahmati, X. Fernando, A. Grami, Steady-state Markov chain analysis for heterogeneous cognitive radio networks, in *IEEE Sarnoff Symposium*, April 2010, pp. 1–5
8. S. Hijazi, B. Natarajan, M. Michelini, Z. Wu, C. Nassar, Flexible spectrum use and better coexistence at the physical layer of future wireless systems via a multicarrier platform. IEEE Wireless Commun. **11**, 8–14 (2004)
9. IEEE standard for information technology, telecommunications and information exchange between systems WRAN specific requirements - Part 22: cognitive wireless RAN Medium Access Control (MAC) and Physical Layer (PHY) specifications, IEEE 802.22a (Amendment to the IEEE Std-802.22-2011(TM)), 2014
10. Standard for wireless LAN in TV white space, IEEE 802.11af, February 2014
11. IEEE 1900 Standards
12. IEEE 802.19: TV white space coexistence methods
13. ETSI, Mobile broadband services in the 2,300 MHz–2,400 MHz frequency band under licensed shared access regime (2013)
14. M. Matinmikko, M. Palola, H. Saarnisaari, M. Heikkila, J. Prokkola, T. Kippola, T. Hanninen, M. Jokinen, S. Yrjola, Cognitive radio trial environment: first live authorized shared access-based spectrum-sharing demonstration. IEEE Veh. Technol. Mag. **8**, 30–37 (2013)
15. IEEE 802.15 Working Group, Wireless medium access control (MAC) and physical layer (PHY) specifications for low-rate wireless personal area networks (WPANs) (2012)
16. H. Celebi, I. Guvenc, S. Gezici, H. Arslan, Cognitive radio systems for spectrum, location, and environment awareness. IEEE Antennas Propag. Mag. **52**(4), 41–61 (2010)
17. A. Ghasemi, E.S. Sousa, Optimization of spectrum sensing for opportunistic spectrum access in cognitive radio networks, in *Proceedings of IEEE International Conference on Consumer Communications and Networking*, Las Vegas, January 2007, pp. 1022–1026
18. S. Hussain, X. Fernando, Closed-form analysis of relay-based cognitive radio networks over Nakagami-m fading channels. IEEE Trans. Veh. Technol. **63**(3), 1193–1203 (2014)
19. S. Hussain, X. Fernando, Performance analysis of relay-based cooperative spectrum sensing in cognitive radio networks over non-identical Nakagami-m channels. IEEE Trans. Commun. **62**(8), 2733–2746 (2014)
20. G. Ding, Q. Wu, Y. Zou, J. Wang, Z. Gao, Joint spectrum sensing and transmit power adaptation in interference-aware cognitive radio networks. IEEE Trans. Emerg. Telecommun. Technol. **25**(2), 231–238 (2014)

21. I.F. Akyildiz, B.F. Lo, R. Balakrishnan, Cooperative spectrum sensing in cognitive radio networks: a survey. Phys. Commun. **4**(1), 40–62 (2011)
22. C. Xing, B. Zhi-Song, W. Wei-Ling, Detection efficiency of cooperative spectrum sensing in cognitive radio network. IEEE Trans. Wirel. Commun. **15**(3), 1–7 (2008)
23. A. Ghasemi, E.S. Sousa, Collaborative spectrum sensing for opportunistic access in fading environments, in *Proceedings of IEEE 1st Symposium on Dynamic Spectrum Access Networks (DySPAN'05)*, Baltimore, November 2005, pp. 131–136
24. Z. Quan, S. Cui, A.H. Sayed, Optimal linear cooperation for spectrum sensing in cognitive radio networks. IEEE J. Sel. Top. Signal Process. **2**(1), 28–40 (2008)
25. D. Cabric, S. Mishra, R. Broderson, Implementation issues in spectrum sensing for cognitive radios, in *Proceedings of Asilomar Conference on Signals, Systems and Computers*, Pacific Grove, November 2004, pp. 772–776
26. G. Ganesan, G. Li, Cooperative spectrum sensing in cognitive radio, part I: two user networks. IEEE Trans. Wirel. Commun. **6**(6), 2204–2213 (2007)
27. G. Ganesan, G. Li, Cooperative spectrum sensing in cognitive radio, part II: multiuser networks. IEEE Trans. Wirel. Commun. **6**(6), 2214–2211 (2007)
28. K. Ben Letaief, W. Zhang, Cooperative communications for cognitive radio networks, in *Proceedings of IEEE*, May 2009, pp. 878–893
29. S.M. Mishra, A. Sahai, R.W. Brodersen, Cooperative sensing among cognitive radios, in *Proceedings IEEE International Conference on Communications (ICC'06)*, Istanbul, June 2006, pp. 1658–1663
30. I.F. Akyildiz, W.-Y. Lee, M.C. Vuran, S. Mohanty, NeXt generation/dynamic spectrum access/cognitive radio wireless networks: a survey. Pervasive Mob. Comput. **50**(12), 2127–2159 (2006)
31. A.S. Zahmati, S. Hussain, X. Fernando, A. Grami, Cognitive radio wireless sensor networks: emerging topics and recent challenges, in *Proceedings IEEE International Conference on Science and Technology for Humanity (TIC-STH'09)*, Toronto, September 2009, pp. 593–596
32. S. Shanker, C. Cordeiro, K. Challapali, Spectrum agile radios: utilization and sensing architecture, in *Proceedings of IEEE International Symposium on Dynamic Spectrum Access Networks, (DySPAN'05)*, Baltimore, November 2005, pp. 160–169
33. J.N. Laneman, N.C. Tse, G.W. Wornell, Cooperative diversity in wireless networks efficient protocols and outage behaviour, IEEE Trans. Inf. Theory **50**(12), 3062–3080 (2004)
34. J. Ma, G. Zhao, Y. Li, Soft combination and detection for cooperative spectrum sensing in cognitive radio networks. IEEE Trans. Wirel. Commun. **7**(11), 4502–4507 (2008)
35. L. Chen, J. Wang, S. Li, Cooperative spectrum sensing with multi-bits local sensing decisions in cognitive radio networks, in *Proceedings IEEE International Conference on Wireless Communications and Networking (WCNC'8)*, New Orleans, April 2008, pp. 570–575
36. S. Haykin, Cognitive radio: brain-empowered wireless communications. Mob. Netw. Appl. **23**(21), 201–220 (2005)
37. G. Ganesan, Y. Li, Cooperative spectrum sensing in cognitive radio, part I: two user networks. IEEE Trans. Wirel. Commun. **6**(6), 2004–2213 (2007)
38. V. Brik, E. Rozner, S. Banarjee, P. Bahl, DSAP: a protocol for coordinated spectrum access, in *Proceedings IEEE 1st Symposium on Dynamic Spectrum Access Networks (DySPAN'05)*, Baltimore, November 2005, pp. 611–614
39. C. Raman, R.D. Yates, N.B. Mandayam, Scheduling variable rate links via a spectrum server, in *Proceedings IEEE 1st Symposium on Dynamic Spectrum Access Networks (DySPAN'05)*, Baltimore, November 2005, pp. 110–118
40. I.F. Akyildiz, J. Xie, S. Mohanty, A survey of mobility management in next-generation all-IP-based wireless system. IEEE Trans. Wirel. Commun. **11**(4), 16–28 (2004)
41. W.-Y. Lee, I.F. Akyildiz, Optimal spectrum sensing framework for cognitive radio networks. IEEE Trans. Wirel. Commun. **7**(10), 3845–3857 (2008)
42. J. Zhao, G. Zheng, H. Yang, Distributed coordination in dynamic spectrum allocation networks, in *Proceedings IEEE 1st Symposium on Dynamic Spectrum Access Networks (DySPAN'05)*, Baltimore, November 2005, pp. 259–268

43. Z. Chair, P.K. Varshney, Optimal data fusion in multiple sensor detection systems. IEEE Trans. Aerosp. Electron. Syst. **22**(1), 98–101 (1986)
44. Q. Zhao, L. Tong, and A. Swami, Decentralized cognitive MAC for dynamic spectrum access, in *Proceedings IEEE 1st Symposium on Dynamic Spectrum Access Networks (DySPAN'05)*, Baltimore, November 2005, pp. 224–232
45. L. Ma, X. Han, C.-C. Shen, Dynamic open spectrum sharing MAC protocol for wireless ad hoc network, in *Proceedings IEEE 1st Symposium on Dynamic Spectrum Access Networks (DySPAN'05)*, Baltimore, November 2005, pp. 203–213
46. K.R. Chowdhury, I.F. Akyildiz, Cognitive wireless mesh networks with dynamic spectrum access. IEEE J. Sel. Areas Commun. **7**(5), 810–836 (2009)
47. W.-Y. Lee, I.F. Akyildiz, Optimal spectrum sensing framework for cognitive radio networks. IEEE Trans. Wirel. Commun. **26**(1), 168–181 (2008)
48. M.K. Simon, M.-S. Alouini, *Digital Communication Over Fading Channels*, 2nd edn. (Wiley, New York, 2005)
49. A. Ghasemi, E.S. Sousa, Interference aggregation in spectrum-sensing cognitive wireless networks. IEEE J. Sel. Top. Sign. Process. **2**(1), 41–56 (2008)
50. S. Shankar, Spectrum agile radios: utilization and sensing architecture, in *Proceedings of IEEE 1st Symposium on Dynamic Spectrum Access Networks (DySPAN'05)*, Baltimore, November 2005, pp. 160–169
51. C. Chou, S. Shanker, H. Kim, K.G. Shin, What and how much to gain by spectrum agility. IEEE J. Sel. Area Commun. **25**(3), 576–588 (2007)
52. H. Kim, K.G. Shin, Efficient discovery of spectrum opportunities with MAC-layer sensing in cognitive radio networks. IEEE Trans. Mob. Comput. **7**(5), 533–545 (2008)
53. K. Siriram, W. Whitt, Characterizing superposition arrival processes in packet multiplexers for voice and data. IEEE J. Sel. Areas Commun. **4**(6), 833–846 (1986)
54. K.T. Kim, S.K. Oh, Cognitive ad-hoc networks under a cellular network with an interference temperature limit, in *Proceedings International Conference on Advanced Communication Technology (ICACT'08)*, Gangwon-Do, February 2008, pp. 879–882
55. F. Xu, Z. Zhou, Y. Ye, Capacity analysis of cognitive UWB networks, in *Proceedings International Conference on Communications and Networking (CHINACOM'08)*, Beijing, August 2008, pp. 752–756
56. J.T. MacDonald, D.R. Ucci, Interference temperature limits of IEEE 802.11 protocol radio channels, in *Proceedings International Conference on Electro/Information Technology*, Chicago, May 2007, pp. 64–69
57. FCC, Notice of Inquiry and Notice of Proposed rulemaking, ET Docket No 03-237 (2003)
58. T. Yucek, H. Arslan, A survey of spectrum sensing algorithms for cognitive radio applications. IEEE Commun. Surv. Tutor. **11**(1), 116–130 (2009)
59. H. Urkowitz, Energy detection of unknown deterministic signals. Proc. IEEE **55**, 523–531 (1967)
60. J. Proakis, *Digital Communications*, 3rd edn. (McGraw Hill, New York, 1995)
61. A. Sahai, N. Hoven, R. Tandra, Some fundamental limits on cognitive radio, in *Proceedings of the Allerton Conference on Communication, Control, and Computing* (2004)
62. W.A. Gardner, C.M. Spencer, Signal interception: performance advantage of cyclic-feature detectors. IEEE Trans. Commun. **40**, 149–159 (1992)
63. M. Oner, F.K. Jondral, On the extraction of the channel allocation information in spectrum pooling system. IEEE J. Sel. Areas Commun. **25**(3), 558–565 (2007)
64. J. Zhu, Z. Xu, F. Wang, B. Huang, B. Zhang, Double threshold energy detection of cooperative spectrum sensing in cognitive radio, in *Proceedings International Conference on Cognitive Radio Oriented Wireless Networks and Communications (CrownCom'08)*, Singapore, May 2008, pp. 1–5
65. F.F. Digham, M.-S. Alouini, M.K. Simon, On the energy detection of unknown signals over fading channels. IEEE Trans. Commun. **55**(1), 3845–3857 (2007)
66. B. Canberk, I.F. Akyildiz, S. Oktug, Primary user activity modeling using first-difference filter clustering and correlation in cognitive radio networks. IEEE/ACM Trans. Netw. **19**(1), 3845–3857 (2011)

67. D.G. Brennan, Linear diversity combining techniques. Proc. IEEE **91**(2), 331–356 (2003)
68. J.N. Laneman, G.W. Wornell, Energy efficient antenna sharing and relaying for wireless networks, in *Proceedings IEEE International Conference on Wireless Communications and Networking (WCNC'00)*, Chicago, October 2000
69. M.O. Hasna, M.-S. Alouini, End-to-end performance of transmission systems with relays over Rayleigh-fading channels. IEEE Trans. Wirel. Commun. **2**(6), 1126–1131 (2003)
70. M. Di Renzo, L. Imbriglio, F. Grazios, F. Santucci, Distributed data fusion over correlated lognormal sensing and reporting channels: application to cognitive radio networks. IEEE Trans. Wirel. Commun. **8**(12), 5813–5821 (2009)
71. M.O. Hasna, M.-S. Alouini, A performance study of dual-hop transmission with fixed gain relays. IEEE Trans. Wirel. Commun. **3**(6), 1963–1968 (2004)
72. M.D. Renzo, L. Imbriglio, F. Grazios, F. Santucci, A comprehensive framework for performance analysis of dual-hop cooperative wireless systems with fixed-gain relays over generalized fading channels. IEEE Trans. Wirel. Commun. **8**(10), 5060–5074 (2009)
73. C. de M. Cordeiro, K.S. Challapali, D. Birru, N.S. Shankar, IEEE 802.22: an introduction to the first wireless standard based on cognitive radios. J. Commun. **1**(1), 43–47 (2006)
74. S.M. Almalfouh, G.L. Stuber, Uplink resource allocation in cognitive radio networks with imperfect spectrum sensing, in *Proceedings IEEE Vehicular Technology Conference (VTC'10)*, September 2010, pp. 1–10
75. M. Thoppian, S. Venkatesan, R. Prakash, R. Chandrasekaran, MAC-layer scheduling in cognitive radio based multi-hop wireless networks, in *Proceedings International Symposium on World of Wireless, Mobile and Multimedia Networks (WoWMoM'06)*, Buffalo-Niagara Falls, June 2006
76. C. Guo, T. Zhang, Z. Zeng, C. Feng, Investigation on spectrum sharing technology based on cognitive radio, in *Proceedings IEEE International Conference on Communications and Networking*, Beijing, October 2006, pp. 1–5
77. T. Eng, N. Kong, L.B. Milstein, Comparison of diversity combining techniques for Rayleigh-fading channels. IEEE Trans. Commun. **44**(9), 1117–1129 (1996)
78. E. Peh, Y.-C. Liang, Optimization for cooperative sensing in cognitive radio networks, in *Proceedings IEEE International Conference on Wireless Communications and Networking (WCNC'7)*, Kowloon, March 2007, pp. 27–32

Chapter 2
Relay-Based Cooperative Spectrum Sensing

In this chapter, the detection accuracy of an AF relay-based CSS approach over non-identical Nakagami-*m* fading channel is investigated. New exact and approximated closed-form expressions are derived for the average detection probability and the average false alarm probability over two diversity combining techniques: MRC scheme and SC scheme. The convergence rate of infinite series that appears in the derived exact closed-form expressions are also investigated and proposed a powerful acceleration algorithm that allows for the series termination with a finite number of terms.

2.1 Spectrum Sensing over Fading Channels

Many urban and vehicular communication systems are subjected to fading caused by multipath propagation due to reflections, refractions and scattering by buildings and other large structures [1]. For primary user detection, flat fading yields the worst case performance since frequency selectivity provides multiple "looks" at the same signal [2]. In a composite fading/shadowing environment, the multipath fading wireless signals may undergo shadowing too. While the multipath fading can be modeled as a Rayleigh, Rice or Nakagami distribution, shadowing process is typically modeled as a Log-normal distribution [2, 3] and the reference therein. In this environment the receiver does not average out the envelope fading due to multipath but rather reacts to the instantaneous composite multipath/shadowed signal [4]. This is often the scenario in congested areas with slow moving pedestrians and vehicles [5]. Therefore, some practical communication channels can be modeled as a multipath fading superimposed on Log-normal shadowing [3]. Due to difficulty of analyzing digital wireless communication systems over composite fading models, the shadowing effect is sometimes neglected in the literature. It is also important to

© The Author(s), under exclusive licence to Springer International Publishing AG, part of Springer Nature 2019
X. Fernando et al., *Cooperative Spectrum Sensing and Resource Allocation Strategies in Cognitive Radio Networks*, SpringerBriefs in Electrical and Computer Engineering, https://doi.org/10.1007/978-3-319-73957-1_2

keep in mind that shadowing is notoriously hard to model accurately and its statistics can vary with the deployment's environment [6].

Performance analysis of wireless communication networks usually involves complicated and cumbersome statistical methods. Several statistical models have been introduced in the literature to describe the fading envelope of the received signal [1, 7–12]. For short term fades, the Rayleigh distribution can be used to characterize the envelope of faded signals over small propagation area while log-normal distribution is used when much wider geographical areas are involved [1]. However, Nakagami-m distribution [7] is the most versatile statistical model which can model a variety of fading environments. Nakagami distribution has a relatively simple analytical form, making it attractive in performance analysis. The fading parameter of the Nakagami-m distribution can describe both severe and weak fading. Therefore, the distribution can be employed in the presence or absence of LOS between the user terminal and the base station [8, 13]. Furthermore, the Nakagami distribution is more flexible and more accurately fits the experimental data for many physical propagation channels than the Log-normal and Rician distributions [1].

The early reported results on communication over Nakagami-fading channels have been obtained based on the use of variants of the CF or the MGF of the sum of gamma variate [1, 13, 14]. This chapter investigates a relay-based CSS strategies that incorporates sensing links, relaying links, diversity combining schemes, and energy detection into a general framework for a comprehensive performance analysis of CR networks with diversity reception. The investigated framework adopts the AF strategy as a relaying mechanism, using an energy detection technique and applying the soft-fusion policy to facilitate the cooperative decision.

2.2 The Spectrum Sensing Models

A centralized CR network with L active secondary users is considered. The cooperative decision is assumed to be made by a fusion center. The secondary users, $\{CR_i\}_{i=1}^{L}$, stand for the system relays and share the same spectrum band, which is originally allocated to the primary users. The channel fading parameters of the ith sensing link and the ith relaying link are denoted by h_{s_i} and h_{r_i}, respectively. We also denote by n_{s_i} and n_{r_i}, the additive Gaussian noise of the sensing and the relaying channels, respectively, which are modeled as white Gaussian random processes that are assumed to be i.i.d with zero mean and variance N_0.

It is assumed that CSI is available for all the relays and also, the fusion center has a full knowledge of CSI, i.e., it can acquire the global knowledge of sensing and relaying channel gains. CSI is one of the main requisites for successful implementation of dynamic cognitive radio protocols. In practice, it is easy to obtain local knowledge of channel gains via a feedback mechanism by using pilot signals transmitted periodically by the fusion center [15, 16]. In order to obtain the CSI of the distant link (i.e. sensing link), CR amplifies a pilot signal received from the primary base station and forwards it to the fusion center [17].

Let x_p denotes the signal transmitted by the primary radio, then the received signal by the ith CR user, y_{s_i}, can be expressed as

$$y_{s_i} = \theta h_{s_i} x_p + n_{s_i} \tag{2.1}$$

where, $\theta = 0$ or 1 denotes the primary user state under two hypotheses: H_o for primary user absence and H_1 for primary user presence. If Y_{s_i} denotes the power of y_{s_i}, then the mean value of Y_{s_i} can be expressed as [18]

$$E\{Y_{s_i}\} = \begin{cases} \sigma_{Y_{s_{i0}}} = N_0 & H_0 \\ \sigma_{Y_{s_{i1}}} = E_i + N_0 = N_0(1 + \overline{\gamma}_{s_i}) & H_1 \end{cases} \tag{2.2}$$

where, $E_i = E\{|h_{s_i} x_p|^2\}$ is the mean value of the signal power as received at the RF front-end of the ith CR receiver and $\overline{\gamma}_{s_i} = E_i/N_0$ is the average SNR associated with the ith sensing link. For local spectrum sensing, Y_{s_i} is compared with a given threshold λ_i to infer the primary state θ. Hence, the false alarm probability, P_{f_i}, and the detection probability, P_{d_i}, can be expressed as

$$P_{f_i} = P\{Y_{s_i} > \lambda_i | H_0\} = \int_{\lambda_i}^{\infty} \left(\frac{m_i}{\sigma_{Y_{s_{i0}}}}\right)^{m_i} \frac{y^{m_i-1}}{\Gamma(m_i)} e^{\left(-\frac{m_i}{\sigma_{Y_{s_{i0}}}}\right)y} dy$$

$$= \frac{\Gamma(m_i, \frac{m_i \lambda_i}{\sigma_{Y_{s_{i0}}}})}{\Gamma(m_i)} = \frac{\Gamma(m_i, \frac{m_i \lambda_i}{N_0})}{\Gamma(m_i)} \tag{2.3}$$

and

$$P_{d_i} = P\{Y_{s_i} > \lambda_i | H_1\} = \int_{\lambda_i}^{\infty} \left(\frac{m_i}{\sigma_{Y_{s_{i1}}}}\right)^{m_i} \frac{y^{m_i-1}}{\Gamma(m_i)} e^{\left(-\frac{m_i}{\sigma_{Y_{s_{i1}}}}\right)y} dy$$

$$= \frac{\Gamma(m_i, \frac{m_i \lambda_i}{\sigma_{Y_{s_{i1}}}})}{\Gamma(m_i)} = \frac{\Gamma(m_i, \frac{m_i \lambda_i}{N_0(1+\overline{\gamma}_{s_i})})}{\Gamma(m_i)} \tag{2.4}$$

where, the above integrals is evaluated with the help of [19, Eq. 3.381.3], m_i is the Nakagami-m fading parameter, $\Gamma(.)$ is the gamma function as defined in [20, Eq.(6.1.1)], and $\Gamma(.,.)$ is the incomplete Gamma function defined in [20, Eq. (6.5.3)].

As can be seen from the expression given in (2.3), the false alarm probability of the individual local detection is not related to the SNR and mainly depends on the decision threshold, λ_i. Therefore, to employ the energy detection, λ_i, must be properly selected to achieve a target detection accuracy. In most research works, e.g. [16, 21, 22], λ_i is selected such that the false alarm probability is bounded by an upper target value. A value of $\overline{P}_f < 0.1$ is recommended as an upper bound by IEEE 802.22 standard [23].

2.2.1 Single-Relay System

According to the AF relaying strategy, CR users are allowed to amplify the received signal and directly relay to the fusion center. We assume that every CR relay has a maximum power constraint, P_i. Hence, it measures the average received signal power and scales it appropriately so that the power constraint is satisfied. Accordingly, the signal received at the fusion center takes the form

$$y_i = h_{r_i}(\sqrt{A_i}(\theta h_{s_i} x_p + n_{s_i})) + n_{r_i} = \theta \sqrt{A_i} h_{r_i} h_{s_i} x_p + \sqrt{A_i} h_{r_i} n_{s_i} + n_{r_i} \tag{2.5}$$

where, A_i denotes the dimensionless amplification factor of the ith CR relay. According to the maximum power constraint, A_i is selected as [24]

$$A_i = \frac{P_i}{E_i + N_0}. \tag{2.6}$$

The second term in Eq. (2.5) clearly indicates that the noise in the dual-hop system cannot be independent of the channel statistics. Accordingly, the propagation characteristics of the second hop must be considered when computing the false alarm probability and the detection probability.

In order to accommodate the two hypotheses H_0 and H_1, we define $g_i = |h_{r_i}|^2$ as the instantaneous channel gain of the ith relaying link and $\overline{g}_i \triangleq E\{|h_{r_i}|^2\}$ as the expected value of g_i. Since h_{r_i} follows a Nakagami-m distribution, it is easy to verify that g_i follows a gamma distribution given by

$$f_{G_i}(g) = \left(\frac{m_i}{\overline{g}_i}\right)^{m_i} \frac{g^{m_i-1}}{\Gamma(m_i)} e^{\left(-\frac{m_i}{\overline{g}_i}\right)g}, \quad g \geq 0. \tag{2.7}$$

Let Y_i denotes the power of the relayed signal y_i, then from (2.5), the mean value of Y_i for a given g can be expressed as

$$E\{Y_i|g\} = \begin{cases} \sigma_{Y_{i_0}} = N_0(1 + A_i g) & H_0 \\ \sigma_{Y_{i_1}} = N_0(1 + (1 + \overline{\gamma}_{s_i})A_i g) & H_1 \end{cases} \tag{2.8}$$

Therefore, the probability of the false alarm for a given g can be evaluated as

$$\overline{P}_{f_i|g}^{AF} = P\{Y_i > \lambda | H_0, g\}$$

$$= \int_\lambda^\infty \left(\frac{m_i}{\sigma_{Y_{i0}}}\right)^{m_i} \frac{y^{m_i-1}}{\Gamma(m_i)} e^{\left(-\frac{m_i}{\sigma_{Y_{i0}}}\right)y} dy$$

$$= \frac{1}{\Gamma(m_i)} \Gamma\left(m_i, \frac{m_i\lambda}{N_0(1+A_ig)}\right) \tag{2.9}$$

where, the integral in (2.9) is evaluated with the help of [19, Eq. 3.383.5] and λ denotes the decision threshold used by the fusion center to infer the primary state θ. Now, we remove the condition on g and compute an average false alarm probability $\overline{P}_{f_i}$ by integrating over the PDF of the channel gain given in (2.7) as follows

$$\overline{P}_{f_i}^{AF} = \int_0^\infty P\{Y_i > \lambda | H_0, g\} f_{G_i}(g) dg$$

$$= \int_0^\infty \frac{1}{\Gamma(m_i)} \Gamma\left(m_i, \frac{m_i\lambda}{N_0(1+A_ig)}\right) f_{G_i}(g) dg. \tag{2.10}$$

Evaluating the integral in (2.10) as described in Appendix A.1, $\overline{P}_{f_i}^{AF}$ can be expressed mathematically as

$$\overline{P}_{f_i}^{AF} = \left(\frac{\beta_i}{A_i}\right)^{m_i} \sum_{q=0}^{m_i-1} \frac{1}{q!} \left(\frac{m_i\lambda}{N_0}\right)^q \sum_{n=0}^{\infty} (-1)^n b_n U\left(m_i; m_i+1-q-n; \frac{\beta_i}{A_i}\right) \tag{2.11}$$

where, $\beta_i = m_i/\overline{g}_i$, $b_n = \frac{1}{n!}\left(\frac{m_i\lambda}{N_0}\right)^n$, and $U(.;.;.)$ is the confluent hypergeometric function of the second kind defined in [20, Eq. 13.1.3]. Similarly, the average detection probability $\overline{P}_{d_i}$ can be obtained as

$$\overline{P}_{d_i}^{AF} = \left(\frac{\beta_i}{(1+\overline{\gamma}_{s_i})A_i}\right)^{m_i} \sum_{q=0}^{m_i-1} \frac{1}{q!}\left(\frac{m_i\lambda}{N_0}\right)^q \sum_{n=0}^{\infty}(-1)^n b_n$$

$$\times U\left(m_i; m_i+1-q-n; \frac{\beta_i}{(1+\overline{\gamma}_{s_i})A_i}\right). \tag{2.12}$$

2.2.2   Multi-Relay CSS System

Selection Combining Scheme

In the performance analysis of cooperative diversity techniques, the statistic of the channel with maximum gain is often important [4]. The selection combining technique is often used to identify this channel among all the diversity branches. It is worth noting that SC technique can be implemented in two different strategies

[25]. In the first strategy, the combiner selects the relaying branch with the highest SNR, γ_r, while in the second strategy, the relay with the highest $\min(\gamma_s, \gamma_r)$ is selected. In our CSS model, we employ the first strategy, however, instead of the SNR, the channel gain will be used to accommodate the two hypotheses, H_0 and H_1. Therefore, for L inputs, the output of the SC receiver is expressed as: $g_{max} = max(g_1, g_2, \cdots, g_L)$. If Y denotes the signal power at the output of the combiner then the mean value of Y for a given g can be expressed as

$$E\{Y|g\} = \begin{cases} \sigma_{Y_0} = N_0(1 + A_{SC}g) & H_0 \\ \sigma_{Y_1} = N_0[1 + (1 + \overline{\gamma}_{ssc})A_{SC}g] & H_1 \end{cases} \tag{2.13}$$

where, $A_{SC} \in \{A_i\}_{i=1}^L$ and $\overline{\gamma}_{ssc} \in \{\overline{\gamma}_{s_i}\}_{i=1}^L$ denote the amplification factor and the average SNR associated with the selected relay, respectively. The average false alarm probability, $\overline{P}_{fsc}$ is evaluated by averaging over the PDF of g_{max} as follows

$$\overline{P}_{fsc} = \int_0^\infty P\{Y > \lambda | H_0, g\} f_{G_{max}}(g) dg$$

$$= \int_0^\infty \frac{1}{\Gamma(m_{SC})} \Gamma(m_{SC}, \frac{m_{SC}\lambda}{N_0(1 + A_{SC}g)}) f_{G_{max}}(g) dg \tag{2.14}$$

where, $m_{SC} \in \{m_i\}_{i=1}^L$ denotes the Nakagami-m fading parameter of the link associated with the selected relay. With the help of Appendix A.2, $\overline{P}_{fsc}$ can be expressed mathematically as follows

$$\overline{P}_{fsc} = \sum_{k=0}^{L-1} \frac{(-1)^k}{k!} \sum_{j=1}^L \frac{\beta_j^{m_j}}{\Gamma(m_j)} \sum_{n_1=1}^L \cdots \sum_{\substack{n_k=1 \\ n_1 \neq n_2 \neq \cdots n_k \neq j}}^L \sum_{l_1=0}^{m_{n_1}-1}$$

$$\cdots \sum_{l_k=0}^{m_{n_k}-1} \left(\prod_{t=1}^k \frac{\beta_{n_t}^{l_t}}{l_t!} \right) \sum_{q=0}^{m_j-1} \frac{1}{q!} \left(\frac{m_{SC}\lambda}{N_0} \right)^q$$

$$\times \frac{\Gamma(v_{kj})}{A_{SC}^{v_{kj}}} \sum_{n=0}^\infty (-1)^n b_n U(v_{kj}; v_{kj} + 1 - q - n; \frac{\mu_{kj}}{A_{SC}}) \tag{2.15}$$

where, $b_n = \frac{1}{n!} \left(\frac{m_{SC}\lambda}{N_0} \right)^n$, $v_{kj} = \sum_{t=1}^k l_t + m_j$, and $\mu_{kj} = \sum_{t=1}^k \beta_{n_t} + \beta_j$. Similarly

$$\overline{P}_{dsc} = \sum_{k=0}^{L-1} \frac{(-1)^k}{k!} \sum_{j=1}^L \frac{\beta_j^{m_j}}{\Gamma(m_j)} \sum_{n_1=1}^L \cdots \sum_{\substack{n_k=1 \\ n_1 \neq n_2 \neq \cdots n_k \neq j}}^L \sum_{l_1=0}^{m_{n_1}-1}$$

$$\cdots \sum_{l_k=0}^{m_{n_k}-1} \left(\prod_{t=1}^{k} \frac{\beta_{n_t}^{l_t}}{l_t!} \right) \sum_{q=0}^{m_j-1} \frac{1}{q!} \left(\frac{m_{SC}\lambda}{N_0} \right)^q \frac{\Gamma(v_{kj})}{((1+\overline{\gamma}_{ssc})A_{SC})^{v_{kj}}}$$

$$\times \sum_{n=0}^{\infty} (-1)^n b_n U(v_{kj}; v_{kj} + 1 - q - n; \frac{\mu_{kj}}{(1+\overline{\gamma}_{ssc})A_{SC}}). \quad (2.16)$$

As a by-product of the above result, $\overline{P}_{fsc}$ and $\overline{P}_{dsc}$ over Rayleigh channels, $\overline{P}_{fsc_{Ray}}$ and $\overline{P}_{dsc_{Ray}}$, respectively, can be obtained by setting $m_j = 1$ and $m_{SC} = 1$ in (2.15) and (2.16)[1]

$$\overline{P}_{fsc_{Ray}} = \sum_{k=0}^{L-1} \frac{(-1)^k}{k!} \sum_{j=1}^{L} \frac{1}{A_{SC}\overline{g}_j} \underbrace{\sum_{n_1=1}^{L} \cdots \sum_{n_k=1}^{L}}_{n_1 \neq n_2 \neq \cdots n_k \neq j} \sum_{n=0}^{\infty} (-1)^n \tilde{b}_n U(1; 2-n; \frac{\tilde{\mu}_{kj}}{A_{SC}})$$

$$(2.17)$$

and

$$\overline{P}_{dsc_{Ray}} = \sum_{k=0}^{L-1} \frac{(-1)^k}{k!} \sum_{j=1}^{L} \frac{1}{(1+\overline{\gamma}_{ssc})A_{SC}\overline{g}_j} \underbrace{\sum_{n_1=1}^{L} \cdots \sum_{n_k=1}^{L}}_{n_1 \neq n_2 \neq \cdots n_k \neq j} \sum_{n=0}^{\infty} (-1)^n \tilde{b}_n$$

$$\times U(1; 2-n; \frac{\tilde{\mu}_{kj}}{(1+\overline{\gamma}_{ssc})A_{SC}}) \quad (2.18)$$

where, $\tilde{b}_n = \frac{1}{n!} \left(\frac{\lambda}{N_0} \right)^n$ and $\tilde{\mu}_{kj} = \sum_{t=1}^{k} \left(\frac{1}{\overline{g}_{n_t}} \right) + \left(\frac{1}{\overline{g}_j} \right)$.

Over i.i.d Nakagami-m channels $\overline{P}_{fsc}$ and $\overline{P}_{dsc}$ are expressed as follows

$$\overline{P}_{fsc} = \frac{\beta^m}{\Gamma(m)} \sum_{k=0}^{L-1} \frac{(-1)^k}{k!} \sum_{j=1}^{L} \underbrace{\sum_{n_1=1}^{L} \cdots \sum_{n_k=1}^{L}}_{n_1 \neq n_2 \neq \cdots n_k \neq j} \sum_{l_0=0}^{m-1} \cdots \sum_{l_k=0}^{m-1} \left(\prod_{t=1}^{k} \frac{\beta^{l_t}}{l_t!} \right) \sum_{q=0}^{m-1} \frac{1}{q!} \left(\frac{m\lambda}{N_0} \right)^q$$

$$\frac{\Gamma(v_k)}{A^{v_k}} \sum_{n=0}^{\infty} (-1)^n b_n U(v_k; v_k + 1 - q - n; \frac{\mu_k}{A}) \quad (2.19)$$

and

---

[1]m=1 is a special case of Nakagami-m fading referred to be as a Rayleigh fading

$$\overline{P}_{dSC} = \frac{\beta^m}{\Gamma(m)} \sum_{k=0}^{L-1} \frac{(-1)^k}{k!} \underbrace{\sum_{j=1}^{L} \sum_{n_1=1}^{L} \cdots \sum_{n_k=1}^{L}}_{n_1 \neq n_2 \neq \cdots n_k \neq j} \sum_{l_1=0}^{m-1} \cdots \sum_{l_k=0}^{m-1} \left(\prod_{t=1}^{k} \frac{\beta^{l_t}}{l_t!} \right) \sum_{q=0}^{m-1} \frac{1}{q!} \left(\frac{m\lambda}{N_0} \right)^q$$

$$\frac{\Gamma(v_k)}{((1+\overline{\gamma}_s)A)^{v_k}} \sum_{n=0}^{\infty} (-1)^n b_n U(v_k; v_k + 1 - q - n; \frac{\mu_k}{(1+\overline{\gamma}_s)A}). \qquad (2.20)$$

In this case, it is important to mention that $v_k = \sum_{t=1}^{k} l_t + m$ and $\mu_k = (k+1)\beta$.

MRC Scheme

Many of performance analysis problems require determination of statistics of the sum of the squared envelopes of the faded signals over several diversity paths which can be achieved through MRC technique. For L inputs, The output of the MRC receiver can be expressed as

$$y = \sum_{i=1}^{L} \sqrt{A_i} h_{r_i} (\theta h_{s_i} x_p + n_{s_i}) + n_{r_i} = \theta \sum_{i=1}^{L} \sqrt{A_i} h_{r_i} h_{s_i} x_P + \sum_{i=1}^{L} \sqrt{A_i} h_{r_i} n_{s_i} + n_{r_i}.$$

$$(2.21)$$

If Y denotes the power at the output of the MRC receiver, then for given g_i's, the mean value of Y can be expressed as

$$E\{Y|g_i{}'s\} = \begin{cases} \sigma_{Y_0} = N_0(1 + \sum_{i=1}^{L} A_i g_i) & H_0 \\ \sigma_{Y_1} = N_0(1 + \sum_{i=1}^{L} (1 + \overline{\gamma}_{s_i}) A_i g_i) & H_1. \end{cases} \qquad (2.22)$$

Therefore, the average false alarm probability for the MRC case is given by

$$\overline{P}_{f_{MRC}} = P\{Y \geq \lambda|H_0\} = \int_0^{\infty} \cdots \int_0^{\infty} P\{Y \geq \lambda|H_0, g_1, \cdots g_L\}$$

$$\cdot f(g_1|H_0) \cdots f(g_L|H_0) dg_1 \cdots dg_L$$

$$= \int_0^{\infty} \cdots \int_0^{\infty} \frac{1}{\Gamma(m_j)} \Gamma\left(m_j, \frac{m_j \lambda}{N_0(1 + \sum_{i=1}^{L} A_i g_i)}\right)$$

$$\cdot f(g_1|H_0) \cdots f(g_L|H_0) dg_1 \cdots dg_L. \qquad (2.23)$$

The expression of $\overline{P}_{f_{MRC}}$ in (2.23) cannot be evaluated straightforward. However, we can rewrite this expression as the expectation over g_i's such that

$$\overline{P}_{f_{MRC}} = E_{g_1, \cdots, g_L} \{P\{Y \geq \lambda|H_0, g_1, \cdots g_L\}\} = E_{g_1, \cdots, g_L} \left\{ \frac{1}{\Gamma(m_j)} \Gamma\left(m_j, \frac{m_j \lambda}{N_0(1 + \sum_{i=1}^{L} A_i g_i)}\right) \right\}.$$

$$(2.24)$$

To simplify the computations of the above expectation, we define the following random variable

$$R = \sum_{i=1}^{L} A_i g_i. \qquad (2.25)$$

Since each g_i follows gamma distribution, the random variable, R, is a sum of mutually independent gamma variates. Thus the expression of $\overline{P}_{f_{MRC}}$ in (2.24) can be expressed as

$$\overline{P}_{f_{MRC}} = \int_0^{\infty} \frac{1}{\Gamma(m_j)} \Gamma(m_j, \frac{m_j \lambda}{N_0(1+r)}) f_R(r) dr \qquad (2.26)$$

where, $f_R(r)$ is the PDF of the random variable R. The above integral can be evaluated as described in Appendix A.3, yielding

$$\overline{P}_{f_{MRC}} = \left[\prod_{j=1}^{L} \left(-\frac{\beta_j}{A_j}\right)^{m_j} \right] \sum_{j=1}^{L} \sum_{v=1}^{m_j} (-1)^v b_{jv} \sum_{q=0}^{m_j-1} \frac{1}{q!} \left(\frac{m_j \lambda}{N_0}\right)^q \sum_{n=0}^{\infty} (-1)^n b_n U(v; v+1-q-n; \frac{\beta_j}{A_j}) \qquad (2.27)$$

Similarly, the average detection probability can be expressed as

$$\overline{P}_{d_{MRC}} = \left[\prod_{j=1}^{L} \left(-\frac{\beta_j}{(1+\overline{\gamma}_{s_j})A_j}\right)^{m_j} \right] \sum_{j=1}^{L} \sum_{v=1}^{m_j} (-1)^v b_{jv} \sum_{q=0}^{m_j-1} \frac{1}{q!} \left(\frac{m_j \lambda}{N_0}\right)^q$$
$$\sum_{n=0}^{\infty} (-1)^n b_n U(v; v+1-q-n; \frac{\beta_j}{(1+\overline{\gamma}_{s_j})A_j}) \qquad (2.28)$$

where

$$b_{jv} = \left[\sum_{j_1=0}^{m_j-v-1} \binom{m_j-v-1}{j_1} B_j^{m_j-v-1-j_1} \sum_{j_2=0}^{j_1-1} \binom{j_1-1}{j_1} \right.$$
$$\left. \times B_j^{j_1-1-j_2} \sum_{j_3=0}^{j_2-1} \binom{j_2-1}{j_1} B_j^{j_2-1-j_3} \cdots \right] \frac{\Delta_j}{(m_j-v)!}$$

where

$$\Delta_j = \prod_{\substack{i=1 \\ i \neq j}}^{L} (\frac{\beta_j}{A_j} - \frac{\beta_i}{A_i})^{m_i} \qquad (2.29)$$

and

$$B_j^t = (-1)^{t+1} t! \sum_{\substack{i=1 \\ i \neq j}}^{L} m_j \left(\frac{\beta_j}{A_j} - \frac{\beta_i}{A_i} \right)^{-(t+1)}. \tag{2.30}$$

Over Rayleigh fading channels, $\overline{P}_{f_{MRC_{Ray}}}$ and $\overline{P}_{d_{MRC_{Ray}}}$ are evaluated by setting $m_j = 1$ in (2.27) and (2.28)

$$\overline{P}_{f_{MRC_{Ray}}} = (-1)^{L+1} \left[\prod_{j=1}^{L} \frac{1}{A_j \overline{g}_j} \right] \sum_{j=1}^{L} \tilde{\Delta}_j \sum_{n=0}^{\infty} (-1)^n \tilde{b}_n U(1; 2 - n; \frac{1}{A_j \overline{g}_j}) \tag{2.31}$$

and

$$\overline{P}_{d_{MRC_{Ray}}} = (-1)^{L+1} \left[\prod_{j=1}^{L} \frac{1}{(1 + \overline{\gamma}_{s_j}) A_j \overline{g}_j} \right] \sum_{j=1}^{L} \tilde{\Delta}_j \sum_{n=0}^{\infty} (-1)^n \tilde{b}_n$$

$$\times U(1; 2 - n; \frac{1}{(1 + \overline{\gamma}_{s_j}) A_j \overline{g}_j}) \tag{2.32}$$

where, $\tilde{\Delta}_j = \prod_{\substack{i=1 \\ i \neq j}}^{L} \left(\frac{1}{A_j \overline{g}_j} - \frac{1}{A_i \overline{g}_i} \right)$. If the diversity paths have i.i.d Nakagami-m fading, then the density function of R is a gamma density given by

$$f_R(r) = \frac{\beta^{Lm}}{A^{Lm} \Gamma(Lm)} r^{Lm-1} e^{-\frac{\beta}{A} r} \tag{2.33}$$

Therefore, the average false alarm probability $\overline{P}_f$ and the average detection probability $\overline{P}_d$ can be expressed as

$$\overline{P}_{f_{MRC}} = \frac{\beta^{Lm}}{A^{Lm}} \sum_{q=0}^{m-1} \frac{1}{q!} \left(\frac{m\lambda}{N_0} \right)^q \sum_{n=0}^{\infty} (-1)^n b_n U(Lm; Lm + 1 - q - n; \frac{\beta}{A}) \tag{2.34}$$

$$\overline{P}_{d_{MRC}} = \frac{\beta^{Lm}}{((1 + \overline{\gamma}_s) A)^{Lm}} \sum_{q=0}^{m-1} \frac{1}{q!} \left(\frac{m\lambda}{N_0} \right)^q \sum_{n=0}^{\infty} (-1)^n b_n$$

$$\times U(Lm; Lm + 1 - q - n; \frac{\beta}{(1 + \overline{\gamma}_s) A}). \tag{2.35}$$

2.3 Convergence Acceleration

Theoretically, the infinite series $\sum_{n=0}^{\infty}(-1)^n b_n$ that appears in the exact closed-form expressions converges as $n \to \infty$. We observed that this series converges for relatively small values of n when λ is small and, therefore, the infinite-series expressions can be accurately computed. However, when λ becomes large ($\lambda \to \infty$), $\overline{P}_d$ as well as $\overline{P}_f$ tend to be very small (for instance, their values can be less than 10^{-3}). Hence, to accurately compute $\overline{P}_d$ and $\overline{P}_f$, a large number of terms need to be evaluated. Specifically, the sum of the infinite series requires the evaluation of b_n over a large value of n. This shortcoming can be avoided by using a convergence acceleration technique. Series acceleration is used to improve the rate of convergence of infinite-series through sequence transformation algorithms [26]. The technique has the potential to reach the series limit within some accuracy using fewer terms than are required. Sequence transformation generates a new sequence based on a partial sum $S_n = \sum_{k=0}^{n}(-1)^k b_k$ for $n = 0, 1, \cdots, N$. The objective is to estimate $\lim_{n\to\infty} S_n$ by using as few partial sums as possible.

One of the powerful algorithms suitable for this purpose is Wynn's ε-algorithm [27] which is a simple recursive scheme that builds a triangle array called epsilon array in which each term of the sequence is determined by the three previous terms with the initial condition $\varepsilon_{-1}^n = 0$, $\varepsilon_0^n = S_n$

$$\varepsilon_{l+1}^n = \varepsilon_{l-1}^{n+1} + \frac{1}{\varepsilon_l^{n+1} - \varepsilon_l^n}, \qquad l = 0, 1, 2, \cdots \tag{2.36}$$

The algorithm repeatedly applies the recursive expression in (2.36) to estimate the converging point of n while constantly adding adequate terms of $\sum_{n=0}^{\infty}(-1)^n b_n$ to reach the required accuracy. The sequence (ε_l^n) is called the lth column, and its construction can be graphically represented as shown in Fig. 2.1. The even columns converge faster to the limit. In particular, the upsweeping diagonals converge very quickly to the limit. This diagonal sequence (circled on Fig. 2.1) is the result of the ε-algorithm and is called the accelerated sequence.

2.4 Approximated Analysis

Further to the acceleration algorithm, we propose new computational approach to derive the average false alarm and the average detection probabilities. If A_i is assumed to be high enough, an approximated expansion of the incomplete gamma function can be used as follows: for a large A_i, we have $(1 + \overline{\gamma}_{s_i})A_i g_i \gg 1$ and $\Gamma(m_i, \frac{m_i}{N_0(1+(1+\overline{\gamma}_{s_i})A_i g_i)})$ is approximated to $\Gamma(m_i, \frac{m_i}{N_0(1+\overline{\gamma}_{s_i})A_i g_i})$, which can be expanded as $(m_i - 1)! e^{\frac{m_i}{N_0(1+\overline{\gamma}_{s_i})A_i g_i}}$. To show how this approximation alleviates the complexity of using the PDF-approach in deriving $\overline{P}_f$ and $\overline{P}_d$, we first derive

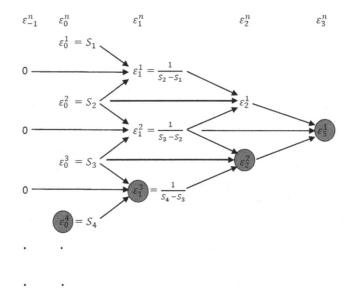

Fig. 2.1 Computation structure of Wynn's ε-algorithm

an approximated close-form expression for $\overline{P}_{dSC}$. Since the average detection probability is evaluated by averaging over the PDF of g_{max}, for $(1+\overline{\gamma}_{SC})A_{SC}g \gg 1$, we then get

$$\overline{P}_{dSC} = \frac{1}{\Gamma(m_{SC})} \int_0^\infty \Gamma\left(m_{SC}, \frac{m_{SC}\lambda}{N_0(1+\overline{\gamma}_{SC})A_{SC}g}\right) f_{G_{max}}(g)dg. \quad (2.37)$$

Substituting (A.5) for $f_{G_{max}}(g)$ and using the series summation $\Gamma(m,x) = (m-1)!e^{-x}\sum_{n=0}^{m-1}\frac{x^n}{n!}$, (2.37) becomes

$$\overline{P}_{dSC} = \sum_{k=0}^{L-1} \frac{(-1)^k}{k!} \sum_{j=1}^{L} \frac{\beta_j^{m_j}}{\Gamma(m_j)} \underbrace{\sum_{n_1=1}^{L} \cdots \sum_{n_k=1}^{L}}_{n_1 \neq n_2 \neq \cdots n_k \neq j} \sum_{l_1=0}^{m_{n_1}-1} \cdots \sum_{l_k=0}^{m_{n_k}-1} \left(\prod_{t=1}^{k} \frac{\beta_{n_t}^{l_t}}{l_t!}\right)$$

$$\sum_{q=0}^{m_i-1} \frac{1}{q!} \left(\frac{m_{SC}\lambda}{N_0(1+\overline{\gamma}_{s_{SC}})A_{SC}}\right)^q \int_0^\infty g^{\nu_{kj}-q-1}e^{-\frac{m_{SC}\lambda}{N_0(1+\overline{\gamma}_{s_{SC}})A_{SC}g}-\mu_{kj}g}dg. \quad (2.38)$$

Using $\int_0^\infty x^{\nu-1}e^{-\frac{\beta}{x}-\gamma x}dx = 2\left(\frac{\beta}{\gamma}\right)^{\frac{\nu}{2}} K_\nu(2\sqrt{\beta\gamma})$ [19, Eq. 3.471.9], to evaluate the integral in (2.38), yields

$$\overline{P}_{d_{SC}} = 2\sum_{k=0}^{L-1}\frac{(-1)^k}{k!}\sum_{j=1}^{L}\frac{\beta_j^{m_j}}{\Gamma(m_j)}\underbrace{\sum_{n_1=1}^{L}\cdots\sum_{n_k=1}^{L}}_{n_1\neq n_2\neq\cdots n_k\neq j}\sum_{l_1=0}^{m_{n_1}-1}\cdots\sum_{l_k=0}^{m_{n_k}-1}\left(\prod_{t=1}^{k}\frac{\beta_{n_t}^{l_t}}{l_t!}\right)\sum_{q=0}^{m_j-1}$$

$$\frac{1}{q!}\left(\frac{m_{SC}\lambda}{N_0(1+\overline{\gamma}_{s_{SC}})A_{SC}}\right)^{\frac{v_{kj}+q}{2}}\left(\frac{1}{\mu_{kj}}\right)^{\frac{v_{kj}-q}{2}}K_{v_{kj}-q}\left(2\sqrt{\frac{m_{SC}\lambda\mu_{kj}}{N_0(1+\overline{\gamma}_{s_{SC}})A_{SC}}}\right)$$

$$(2.39)$$

where, $K_v(.)$ is the v-order modified Bessel function of the second kind defined in [20, Eq. 9.6.22]. It is easily shown that the above expression does not have the infinite-series term, and therefore, it provides a simple computational method to compute the average detection provability.

For independent not identically distributed Rayleigh paths, we obtain

$$\overline{P}_{d_{SC_{Ray}}} = 2\sum_{k=0}^{L-1}\frac{(-1)^k}{k!}\sum_{j=1}^{L}\frac{1}{\overline{g}_j}\underbrace{\sum_{n_1=1}^{L}\cdots\sum_{n_k=1}^{L}}_{n_1\neq n_2\neq\cdots n_k\neq j}\sqrt{\frac{\lambda}{N_0(1+\overline{\gamma}_{s_{SC}})A_{SC}\tilde{\mu}_{kj}}}K_1\left(2\sqrt{\frac{\lambda\tilde{\mu}_{kj}}{N_0(1+\overline{\gamma}_{s_{SC}})A_{SC}}}\right)$$

$$(2.40)$$

and for, i.i.d. Nakagami-m paths

$$\overline{P}_{d_{SC}} = \frac{2\beta^m}{\Gamma(m)}\sum_{k=0}^{L-1}\frac{(-1)^k}{k!}\sum_{j=1}^{L}\underbrace{\sum_{n_1=1}^{L}\cdots\sum_{n_k=1}^{L}}_{n_1\neq n_2\neq\cdots n_k\neq j}\sum_{l_1=0}^{m-1}\cdots\sum_{l_k=0}^{m-1}\left(\prod_{t=1}^{k}\frac{\beta^{l_t}}{l_t!}\right)$$

$$\sum_{q=0}^{m-1}\frac{1}{q!}\left(\frac{m\lambda}{N_0(1+\overline{\gamma}_s)A}\right)^{\frac{v_k+q}{2}}\left(\frac{1}{\mu_k}\right)^{\frac{v_k-q}{2}}K_{v_k-q}\left(2\sqrt{\frac{m\lambda\mu_k}{N_0(1+\overline{\gamma}_s)A}}\right).$$

$$(2.41)$$

The approximation can be used in a similar way to derive the average detection probability of the MRC scheme. Therefore, under independent not identically distributed Nakagami-m paths we have

$$\overline{P}_{d_{MRC}} = 2\left[\prod_{j=1}^{L}\left(-\frac{\beta_j}{(1+\overline{\gamma}_{s_j})A_j}\right)^{m_j}\right]\sum_{j=1}^{L}\sum_{v=1}^{m_j}\frac{(-1)^v b_{jv}}{\Gamma(v)}\sum_{q=0}^{m_j-1}\frac{1}{q!}\left(\frac{m_j\lambda}{N_0}\right)^{\frac{v+q}{2}}$$

$$\times\left(\frac{(1+\overline{\gamma}_{s_j})A_j}{\beta_j}\right)^{\frac{v-q}{2}}K_{v-q}\left(2\sqrt{\frac{m_j\beta_j\lambda}{N_0(1+\overline{\gamma}_{s_j})A_j}}\right).$$

$$(2.42)$$

and

$$\overline{P}_{d_{MRC_{Ray}}} = 2(-1)^{L+1} \left[\prod_{j=1}^{L} \left(-\frac{1}{(1+\overline{\gamma}_{s_j})A_j\overline{g}_j} \right) \right] \sum_{j=1}^{L} b_{j1} \sqrt{\frac{\lambda(1+\overline{\gamma}_{s_j})A_j\overline{g}_j}{N_0}}$$

$$K_1 \left(2\sqrt{\frac{\lambda}{N_0(1+\overline{\gamma}_{s_j})A_j\overline{g}_j}} \right). \tag{2.43}$$

also, for i.i.d. Nakagami-m paths, we have

$$\overline{P}_{d_{MRC}} = \frac{2\beta^{Lm}}{((1+\overline{\gamma}_s)A)^{Lm} \Gamma(Lm)} \sum_{q=0}^{m-1} \frac{1}{q!} \left(\frac{m\lambda}{N_0} \right)^{\frac{Lm+q}{2}} \left(\frac{(1+\overline{\gamma}_s)A}{\beta} \right)^{\frac{Lm-q}{2}}$$

$$\times K_{Lm-q} \left(2\sqrt{\frac{m\beta\lambda}{N_0(1+\overline{\gamma}_s)A}} \right). \tag{2.44}$$

The same approximation approach can also be used to derive the approximated $\overline{P}_{f_{SC}}$ and $\overline{P}_{f_{MRC}}$ under the assumption $A_i g_i \gg 1$.

To validate the accuracy of the approximation, we define

$$I_k = \sum_{q=0}^{m-1} \frac{1}{q!} \left(\frac{m\lambda}{N_0} \right)^q \int_0^\infty (1+(1+\overline{\gamma}_s)Ag)^{-q} g^{v-1} e^{-\frac{m\lambda}{N_0(1+(1+\overline{\gamma}_s)Ag)}} e^{-\mu g} dg. \tag{2.45}$$

For $(1+\overline{\gamma}_s)Ag \gg 1$, I_k in (2.45) can be approximated as

$$I_k \approx \sum_{q=0}^{m-1} \frac{1}{q!} \left(\frac{m\lambda}{N_0(1+\overline{\gamma}_s)A} \right)^q \int_0^\infty g^{v-q-1} e^{-\frac{m\lambda}{N_0(1+\overline{\gamma}_s)Ag}} e^{-\mu g} dg. \tag{2.46}$$

Using [19, Eq. 3.471.9], the above integral is evaluated as follows

$$I_k \approx 2 \left(\frac{m\lambda}{N_0(1+\overline{\gamma}_s)A} \right)^{\frac{v+q}{2}} \left(\frac{1}{\mu} \right)^{\frac{v-q}{2}} K_{v-q} \left(2\sqrt{\frac{m\lambda\mu}{N_0(1+\overline{\gamma}_s)A}} \right) \tag{2.47}$$

It can be seen from the closed-form expressions derived for $\overline{P}_f$ and $\overline{P}_d$ in (2.15), (2.16), and (2.27) that all the finite sums can be calculated exactly and I_k is the only approximated term in these expressions. Therefore, validating the computation accuracy of I_k, inevitably validates the accuracy of the expressions given in (2.39)–(2.44). In fact, validating I_k offers a simple mean to avoid performing multiple simulation tests to validate each of the derived approximated expressions individually.

2.5 Performance Evaluation

To validate the accuracy of the derived closed-form expressions, we assume that L secondary users are deployed in a centralized CR network. The sensing and the relaying channels are assumed to be subjected to independent not identically distributed Nakagami-m fading. An upper bound of $\overline{P}_f < 0.1$ is considered for the selection of decision threshold. This upper bound is recommended in the literature as well as IEEE 802.22 standard [23]. The noise variance N_0 is set to a unity (0 dB).

To determine the number of terms N required to evaluate the infinite-series expressions derived for $\overline{P}_d$ in (2.16) and (2.28), we list in Table 2.1 the minimum number of terms required to evaluate these expressions for the given values of the threshold decision, λ. As can be seen, the infinite series converges with fewer number of terms for small λ and an accuracy up to four decimal points is obtained. However, as λ increases, the infinite series diverges and unbounded values of the average detection probability are observed. Clearly, for large λ a large number of sum up terms is required to evaluate $\overline{P}_d$ using the exact expressions in (2.16) and (2.28) and it is more appropriate to use the approximated expressions in (2.39) and (2.42) due to their computational simplicity.

Using a large number of terms to evaluate the infinite series, $\sum_{n=0}^{\infty}(-1)^n b_n$, significantly slows down the convergence rate. Therefore, when λ becomes large, it is necessary to use the ε-algorithm to improve the convergence rate. Table 2.2 illustrates the number of terms required to evaluate $\overline{P}_d$ with an accuracy of up to four decimal points using the ε-algorithm to accelerate (2.16) and (2.28). We observed that the use of ε-algorithm dramatically reduces the points of convergence N and highly improves the convergence rate.

To validate the accuracy of the approximated expression, we plot in Fig. 2.2, I_k versus λ for $P = 0, 5, 10$, and 20 dB. In this figure, the analytical results obtained from the I_k definition given in (2.47) are compared to Monte Carlo simulations generated over 100,000 iterations to evaluate the integral-form expression given in (2.45). Obviously, for large λ the analytical results match well with the simulation

Table 2.1 Converging point required to evaluate $\overline{P}_d$ for SC and MRC schemes. $L = 3, m = 3$

	Converging point (n)				
	$\lambda = 20$	$\lambda = 40$	$\lambda = 60$	$\lambda = 80$	$\lambda = 100$
SC Scheme, Eq. (2.16)	9	23	42	69	102
MRC Scheme, Eq. (2.28)	5	16	32	50	69

Table 2.2 Converging point required to evaluate (2.16) and (2.28) with ε-algorithm. $L = 3$, $m = 3$

	Converging point (n)				
	$\lambda = 20$	$\lambda = 40$	$\lambda = 60$	$\lambda = 80$	$\lambda = 100$
SC Scheme, Eq. (2.16)	8	12	16	19	22
MRC Scheme, Eq. (2.28)	3	7	10	13	16

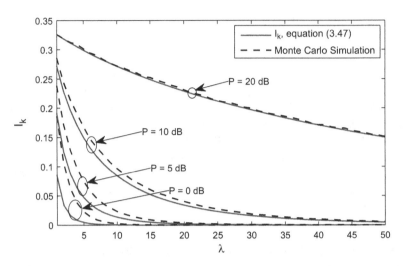

Fig. 2.2 Comparison of analytical results of expression (2.47) with Monte Carlo simulations. $L = 3$, $m = 3$

results and an accuracy of four decimal points is achieved for $\lambda > 15$ with all the power constraint scenarios. For small λ the few number of terms required to evaluate $\sum_{n=0}^{\infty}(-1)^n b_n$ makes the infinite-series expressions more appropriate to compute $\overline{P}_f$ and $\overline{P}_d$ due to their high accuracy. It is also observed that the analytical and the simulation curves become closer to each other as the power constraint increases which comes from the fact that the errors resulting from approximating $N_0(1 + (1 + \overline{\gamma}_{s_i})A_i)$ into $N_0(1 + \overline{\gamma}_{s_i})A_i$ decrease as P increases.

In Fig. 2.3 we compare the analytical results obtained for $\overline{P}_d$ using the exact expressions in (2.16) and (2.28) and the approximated expressions in (2.39) and (2.42) to Monte Carlo simulation results obtained over 100,000 iterations. Generally, the analytical results match well with the simulated ones for both schemes. However, as λ increases, the infinite sum requires the evaluation of $\sum_{n=0}^{\infty}(-1)^n b_n$ over large n. Having an insufficient number of terms to evaluate $\sum_{n=0}^{\infty}(-1)^n b_n$ results in larger convergence errors which explains the slight divergence of the exact curve of the SC scheme from the simulation curve as λ increases. Compared with the MRC scheme, the approximated curve of the SC scheme shows a wider mismatch with the simulated one specifically, at small λ This mismatch is expected as the approximation errors in the case of MRC scheme are less than that of the SC scheme.

In Fig. 2.4, we describe the receiver's performance through its complementary receiver operating characteristic (ROC) curves. The performance of the proposed AF approach is evaluated using the expression given in (2.28) and compared with a DF approach for different channel conditions. For the DF approach, a false alarm probability and a detection probability are computed at each CR user by $P_{f_i} = \Gamma(m_i, \frac{m_i \lambda_i}{N_0})/\Gamma(m_i)$ and $P_{d_i} = \Gamma(m_i, \frac{m_i \lambda_i}{N_0(1+\overline{\gamma}_{s_i})})/\Gamma(m_i)$, respectively.

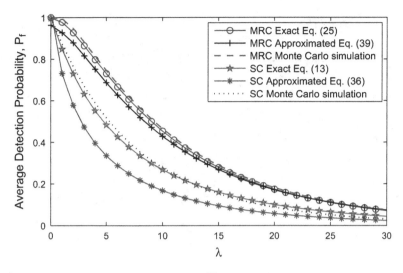

Fig. 2.3 Comparison of analytical and simulated $\overline{P}_d$. $P = 5\,\text{dB}$, $L = 3$, $m = 2$, $\overline{\gamma}_s = 0\,\text{dB}$

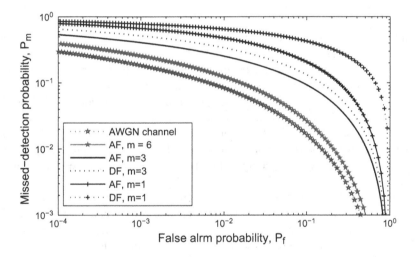

Fig. 2.4 Complementary ROC curves at different relaying channel conditions. $P = 5\,\text{dB}$, $L = 3$, $\overline{\gamma}_s = 0\,\text{dB}$

Where, λ_i is the local decision threshold. The cooperative decision is then computed at the fusion center using a logic-based OR-rule. A similar approach is introduced in [28]. The figure illustrates that the AF approach outperforms the DF approach for the selected values of the fading parameter m. The AWGN case is also plotted for comparison purposes. AWGN channels are frequently assumed in literature to investigate the energy detection performance [26, 29–31]. Clearly, such an assumption overestimates the detection accuracy as can be seen from the plotted curves.

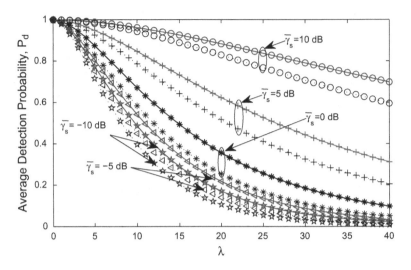

Fig. 2.5 Average detection probability for different level of sensing channel SNR. $P = 5\,\text{dB}$, $L = 3, m = 2$

In Fig. 2.5, we plot the detection probability of the SC scheme (dot-line curves) and the MRC scheme (solid-line curves) for deferent values of the sensing channel average SNR, $\overline{\gamma}_s$. The average detection probability of the SC and MRC schemes are computed using the exact expressions given in (2.16) and (2.28), respectively. There is an obvious improvement in the performance of both combining schemes with each step of 5 dB increase in $\overline{\gamma}_s$ from -10 to 10 dB. The figure shows that the MRC scheme outperforms the SC scheme for all the selected values of $\overline{\gamma}_s$. For low SNR applications, it is more appropriate to the MRC schemes due to the significant difference in its performance compared to the SC scheme as can be seen from the plotted curves.

Figure 2.6 shows that the detection accuracy is significantly improved by increasing the power constraint, P. The average detection probability is plotted for three Nakagami-m fading scenarios with the case $m = 1$ corresponding to Rayleigh fading. For $m = 6$, a value of $\overline{P}_d > 0.9$ is achieved when P increases to 8 dB. This matches the target value of $\overline{P}_d = 0.9$ which is frequently used in literature as a lower bound to the detection probability [23]. This target value is also achieved for other m scenarios but with higher power constraints. The high performance achieved with higher values of the fading parameter, m, refers to the fading severity inversely proportional with m. Furthermore, the MRC scheme shows a better performance compared to the SC scheme, specifically at the low power region. Roughly, a gain of 3 dB is achieved within the range 3–10 dB of the power constraint for $m = 3$ when the combiner is switched from the SC scheme to the MRC scheme, yet the SC performance becomes very close to the MRC performance at higher values of m. Beyond the target value, the curves slow down as $\overline{P}_d$ approaches unity. Thus, there is no need to increase the transmission power to push $\overline{P}_d$ beyond this target

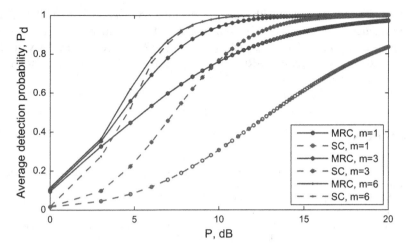

Fig. 2.6 Average detection probability vs. relay maximum transmitting power, P. $L = 3$, $\lambda = 20$, $\overline{\gamma}_s = 0\,\mathrm{dB}$

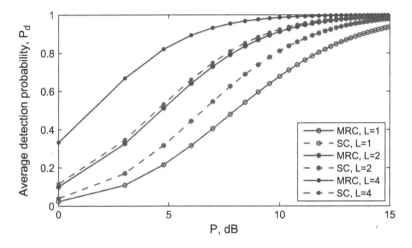

Fig. 2.7 Average detection probability vs. relay power constraint, P. $m = 3$, $\lambda = 20$, $\overline{\gamma}_s = 0\,\mathrm{dB}$

value, since the higher transmission power leads to a higher interference level that may affect the operation of the primary network.

To demonstrate the importance of the diversity reception, we plot $\overline{P}_d$ versus P in Fig. 2.7 for a different number of cooperative users. The expressions of $\overline{P}_d$ given in (2.16) and (2.28) are used with the number of diversity branches varies from 1 to 4. The case of $L = 1$ corresponds to the no-diversity scenario. Observe, that for the no-diversity case, the curve of the MRC scheme coincides with that of the SC schemes. The plotted curves indicate that a higher diversity gain is achieved when the number of cooperative users increases. For instance, with $L = 4$ users, the MRC receiver achieves $\overline{P}_d > 0.9$ with 8 dB less than the no-diversity case while the SC scheme achieves 3 dB gain with the same number of cooperative users. For the dual branch case, a gain of 4 dB is achieved with the MRC scheme and 2 dB with the SC scheme. Furthermore, the MRC scheme outperforms the SC scheme for all the investigated scenarios.

References

1. V.A. Aalo, Performance of maximal-ratio diversity systems in a correlated Nakagami-fading environment. IEEE Trans. Commun. **43**(8), 2360–2369 (1995)
2. T. Rappaport, *Wireless Communications: Principles and Practice*, 2nd edn. (Prentice-Hall, Upper Saddle River, 2002)
3. S. Atapattu, C. Tellambura, H. Jiang, Performance of an energy detector over channels with both multipath fading and shadowing. IEEE Trans. Wirel. Commun. **9**(12), 3662–3670 (2010)
4. M.K. Simon, M.-S. Alouini, *Digital Communication over Fading Channels* (Wiley, New York, 2005)
5. M.S. Alouini, A.J. Goldsmith, A unified approach for calculating the error rates of linearly modulated signals over generalized fading channels, in *Proceedings of IEEE International Conference on Communications (ICC'98)*, June 1998, pp. 1324–1334
6. S.M. Mishra, A. Sahai, R.W. Brodersen, Cooperative sensing among cognitive radios, in *Proceedings IEEE International Conference on Communications, (ICC'06)*, Istanbul, June 2006, pp. 1658–1663
7. M. Nakagami, *The m-Distribution: A General Formula of Intensity Distribution of Rapid Fading*. Statistical Methods in Radio Wave Propagation (Pergamon, New York, 1960)
8. M. Abdel-Hafez, M. Safak, Performance analysis of digital cellular systems in Nakagami fading and correlated shadowing environmental. IEEE Trans. Veh. Tech. **48**(5), 1381–1391 (1999)
9. R. Kwan, C. Leung, General order selection combining for Nakagami and Weibull fading channels. IEEE Trans. Wirel. Commun. **6**(6), 2027–2033 (2007)
10. G.K. Karagiannidis, N.C. Sagias, T.A. Tsiftsis, Closed-form statistics for the sum of squared Nakagami-*m* variates and its applications. IEEE Trans. Commun. **54**(8), 1353–1359 (2004)
11. M.-S. Alouni, A. Abdi, M. Kaveh, Sum of gamma variates and performance of wireless communication systems over Nakagami-fading channels. IEEE Trans. Veh. Tech. **50**(6), 1471–1480 (2001)
12. Q.T. Zhang, Maximal-ratio combining over Nakagami fading channels with an arbitrary ranch covariance matrix. IEEE Trans. Veh. Tech. **48**(4), 1141–1150 (1999)
13. H. Suzuki, A statistical model for the urban radio propagation. IEEE Trans. Commun. **25**, 673–680 (1997)
14. E.K. Al-Hussaini, A.A.M. Al-Bassiouni, Performance of MRC diversity systems for the detection of signals with Nakagami fading. IEEE Trans. Commun. **33**(12), 1315–1319 (1985)
15. X. Qin, R.A. Berry, Distributed approaches for exploiting multiuser diversity in wireless networks. IEEE Trans. Commun. **52**(2), 392–413 (2006)
16. G. Ganesan, G. Li, Cooperative spectrum sensing in cognitive radio, part I: two user networks. IEEE Trans. Wirel. Commun. **6**(6), 2204–2213 (2007)
17. H. Yomo, E. De Carvalho, A CSI estimation method for wireless relay network. IEEE Trans. Commun. **11**(6), 480–482 (2007)
18. S. Hussain, X. Fernando, Closed-form analysis of relay-based cognitive radio networks over Nakagami-m fading channels. IEEE Trans. Veh. Tech. **63**(3), 1193–1203 (2014)
19. I.S. Gradshteyn, I.M. Ryzhik, *Table of Integrals, Series and Products*, 6th edn. (Academic, London, 2000)

20. M. Abramowitz, I.A. Stegun, *Handbook of Mathematical Functions with Formulas, Graphs, and Mathematical Tables*, 9th edn. (Dover, New York, 1972)
21. G. Ganesan, G. Li, Cooperative spectrum sensing in cognitive radio, part II: multiuser networks. IEEE Trans. Wirel. Commun. **6**(6), 2214–2211 (2007)
22. W.-Y. Lee, I.F. Akyildiz, Optimal spectrum sensing framework for cognitive radio networks. IEEE Trans. Wirel. Commun. **7**(10), 3845–3857 (2008)
23. C. Stevenson, G. Chouinard, Z. Lei, W. Hu, S. Shellhammer, W. Caldwell, IEEE 802.22: the first cognitive radio wireless regional area network standard. IEEE Commun. Mag. **47**(1), 130–138 (2009)
24. J.N. Laneman, G.W. Wornell, Energy efficient antenna sharing and relaying for wireless networks, in *Proceedings of IEEE International Conference on Wireless Communications and Networking (WCNC'00)*, Chicago (2000)
25. D.S. Michalopoulos, G.K. Karagiannidis, Performance analysis of single relay selection in Rayleigh fading. IEEE Trans. Wirel. Commun. **7**(10), 3718–3724 (2008)
26. T. Sauter, Computation of irregularly oscillating integrals. Appl. Numer. Math. **35**(3), 245–264 (2000)
27. P. Wynn, Acceleration techniques in numerical analysis, with particular reference to problems in one independent variable. Stichting Mathematisch Centrum. Rekenafdeling (1962), pp. 149–156
28. K. Ben Letaief, W. Zhang, Cooperative communications for cognitive radio networks, in *Proceedings of the IEEE*, May 2009, pp. 878–893
29. Z. Quan, S. Cui, A.H. Sayed, Optimal linear cooperation for spectrum sensing in cognitive radio networks. IEEE J. Sel. Top. Signal Process. **2**(1), 28–40 (2008)
30. E. Peh, Y.-C. Liang, Optimization for cooperative sensing in cognitive radio networks, in *Proceedings of IEEE International Conference on Wireless Communications and Networking (WCNC'7)*, Kowloon, March 2007, pp. 27–32
31. Y.-C. Liang, Y. Zeng, E.C.Y. Peh, A.T. Hoang, Sensing-throughput tradeoff for cognitive radio networks. IEEE Trans. Wirel. Commun. **7**(4), 1325–1337 (2008)

Chapter 3
Taxonomy for the Resource Allocation in CRNs

In this chapter, a taxonomy for the resource allocation (RA) process in CRNs is provided. In particular, the RA problem is categorized first according to the adopted approach which could be centralized, or distributed. Second, based on the network architecture, the RA problem is also classified as infrastructure based or ad-hoc. Third, according to the problem objective, the basic elements of RA in CRNs may include: throughput (sum-rate), power, delay, QoS, fairness etc. Finally, the algorithms are also classified according to the strategy used for solving the problem which can be an optimization technique, heuristic, game theory, graph theory etc. Figure 3.1 provides the adopted taxonomy in detail.

3.1 Resource Allocation Approaches

In the literature, two key approaches are utilized to solve the RA problem: centralized and distributed. Each of these approaches has its merits and demerits. So that, depending on the problem requirement and priorities one of them could be adopted.

1. *Centralized*: The centralized approach mainly based on the existence of a central entity that handles the RA process. This entity could be a base station (BS), eNodeB, or separate node for control purposes (server). In the centralized RA or scheduling algorithms, the central node gathers the measurements and information from the whole network, transmits control information to different users and nodes to coordinate their access, and takes the final decision. This approach has been widely investigated in literature (e.g., [1–7]). It was shown that the centralized schemes have several advantages. Due to the global view of the whole network, this approach is able to obtain the optimal solution of

X. Fernando et al., *Cooperative Spectrum Sensing and Resource Allocation Strategies in Cognitive Radio Networks*, SpringerBriefs in Electrical and Computer Engineering, https://doi.org/10.1007/978-3-319-73957-1_3

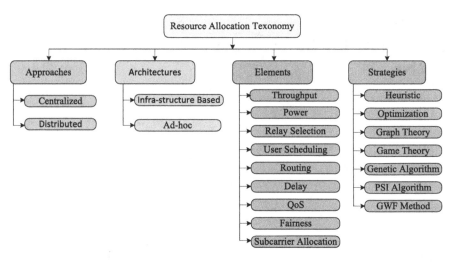

Fig. 3.1 Taxonomy for the resource allocation in CRNs

a desired performance metric (e.g., maximize the network sum rate, spectrum efficiency, etc.). Also, it can optimally minimize the network interference and attain an overall better network performance. Moreover, it can easily achieve optimal fairness since all the network information is available at one node. This can be done by regulating greedy nodes that devour large portion of the resources on the cost of other weak nodes. Moreover, centralized controllers can control priorities more efficiently. In contrast, there are some disadvantages that limit the practicality of the centralized approach. In particular, centralized schemes create large signaling overhead which results in a wasteful utilization of the resources (e.g., bandwidth, power, etc.). Also, due to the adopted nature which directs to make the RA process of the whole network based strongly on a single entity (central node), the failure of this central node can cause severe effects on the network performance. In a failure case, which may arise from power failure or crashes, each node in the network will independently make the RA decision. Consequently, the network performance could be significantly deteriorated. Furthermore, contention and unfairness situation could be attained.

2. *Distributed*: In the distributed approach, there is no central entity that manages the scheduling process. Instead, each node (or user) takes its decision autonomously or via a limited cooperation by exchanging neighboring information. In particular, each node carries out some measurements and calculations that are enough to reach the decision or should be shared among its neighbors to take the decision. Several distributed approaches have been taken into account in many researches (e.g., [8–11]). Flexibility is a major advantage of this approach over the centralized one. Distributed schemes quickly adapt to wireless environment variations. This is due to the fact that only the nodes in the area of the variation are affected and require to amend their calculations. Therefore, there is no need to replicate the overall RA process, but only the

affected part. This definitely involves the fast recovery feature which is crucial, for example, in emergency situations. This contradicts the centralized case, where any change affects the whole process and delays the decisions. Thus, distributed approaches present more robust communications. Another advantage is that distributed approaches decrease the amount of overhead and delay of switching information as a result of being only limited to a small neighboring area. In compare, distributed approaches cannot acquire the optimal solution for the desired performance metric due to the lack of information at each node. Moreover, only local fairness between neighboring nodes can be attained. Finally, distributed approaches are more appropriate for light load networks but for high load networks a centralized approach can achieve better performance.

3.2 Architecture

1. *Infra-Structure Based*: This model incorporates the existence of a base station and multiple users connected to it. The network can be driven either in downlink or uplink modes, and offers one-hop communications. Generally, centralized techniques are utilized for this model to employ the central entity. The users observations and measurements are fed back to the central point to take the decisions. Then, the users configure their parameters according to the central decision. This model is useful in case of setting up a permanent network. In CRNs, using proper RA strategies the base station can significantly help in protecting the primary network from secondary transmissions. Cognitive radio cellular networks (CRCN) [12, 13], and IEEE 802.22 wireless regional area network (WRAN) [14, 15] are examples of such networks. Although most of the strategies in literature for this model are centralized, distributed schemes can also be supported [16].

2. *Ad-Hoc*: In the ad-hoc model, also called infrastructureless model, the communication is performed directly between the CR nodes without the support of a central controller or a base station. Unlike infra-structure based model where the nodes communicate with the base station via single hop, the communication between the nodes of an ad-hoc network could be single-hop or multi-hop communications. In a multi-hop network, routing is an essential function to find the best route for data transmission. Indeed, the use of distributed techniques is more appropriate for this network model as the ad-hoc networks are distributed by nature [17, 18]. This includes extra challenges on the communicating nodes of the ad-hoc networks compared to those of infrastructure-based since the nodes have to coordinate their access themselves. As a result, the operational load increases on the CR nodes whose resources, such as power, are limited which enforces a critical demand for computationally efficient methods.

Ad-hoc networks have several advantages. For instance, ad-hoc networks are easier to set up because the nodes are connected directly without an access point.

Besides the ease of deployment, ad-hoc networks decreased dependence on infra-structure allows for fast integration with already existing infra-structure networks. This makes wireless ad-hoc networks applicable and useful in many situations such as crisis response, military environments and sensor networks [17]. In CRNs, ad-hoc networks were broadly adopted (e.g., [18, 19]). Dynamic network topology, distributed multi-hop architecture, and time and location varying spectrum avail-ability are some prime factors of CR ad-hoc networks (CRAHNs) [18]. The specific functions associated with CRAHNs were discussed in [20]. The authors described that these functions are highly dependent on cooperation among the users due to the lack of a central coordinator. Therefore, RA is done based on local observations of each user and by exchanging information with the neighboring nodes.

3.3 Basic Elements or Objectives

There are several elements or objectives that could be targeted by the RA problem. These elements may be incorporated as the optimization principle objective or as a constraint that should hold.

1. *Throughput*: Maximizing the network throughput is a very familiar criterion that is generally targeted in the RA problem. The problem could consider the individual user throughput maximization or the sum-rate of the aggregate network. Also, some efforts consider maximizing the sum-rate of both primary and secondary networks as a whole.
2. *Power allocation*: Power allocation for CRNs is a crucial task for better interference management. In case of CRNs, the efficient power allocation is more challenging than non-cognitive wireless network. Traditional power allocation schemes for non-cognitive networks are not applicable to CRNs as these schemes may cause unacceptable interference to the primary network. In CRNs, power allocation is performed under the constraint of acceptable interference to the PUs.
3. *Relay assignment/selection*: The use of relays in a CRN can benefit in two ways. First it can increase the transmission rate and, secondly the use of relays can reduce the overall transmission power of the systems. The use of multiple relays simultaneously can further increase the performance of a cognitive radio network. A well designed multiple relay assignment and power allocation scheme can be helpful in two ways. It reduces the interference induced to the primary users in multiuser CRNs and increases the connectivity of the wireless network. In a multiple relay system, if any relay is dead or in deep fade the receiver can still get data from other relays.
4. *User scheduling*: In multiuser CRNs, due to resource limitations and interference constraints, user scheduling in intelligent way can achieve high throughput. User scheduling schemes select the best group of users at each time slot to maximize

the total throughput. The complexity of an exhaustive search for user scheduling increases exponentially with the number of users. For example, if U is the total number of users, then the number of possible ways of scheduling/selecting u users is $\binom{U}{u}$. Enumerating all possible combinations to find the one that gives the best performance is computationally inefficient. Due to the high computational complexity of the optimal selection (e.g., exhaustive search algorithm), efficient user scheduling in CRNs is an active area research.

5. *Routing*: Most of the research on CRNs to date has focused on one or two-hop scenarios. With the advancement on ad hoc networks, recently, researchers have started to realize the importance and potential of multi-hop CRNs. To get the benefits of multi-hop transmission, new challenges must be addressed and solved. In particular, efficient routing techniques and solutions must be integrated into the ad hoc CRNs.

6. *QoS*: QoS is a general term used for many user satisfaction related requirements. It comprises response time, throughput loss, rate requirements, outage and blocking probabilities. The main aim of QoS in CRNs is to guarantee a minimum rate, reduction in latency jitter and packet errors.

7. *Delay*: Delay is an important metric in any wireless network especially for real-time applications such as voice and multimedia. Delay performance in CRNs has important engineering implications, e.g., it can be used to characterize the number of SUs that can be supported under a given delay constraint. Queuing delay and transmission delay need to be analyzed in order to characterize the delay performance, which is still an unexplored area of research.

8. *Subcarrier allocation*: Subcarrier allocation and pairing play a significant role in future CRNs that employs orthogonal frequency division multiplexing (OFDM) in physical layer. One can increase the throughput of CRNs with the intelligent utilization of subcarriers.

Figure 3.2 shows the generic resource allocation problem in CRNs. A generic RA design structure consists of four portions: (1) inputs; (2) outputs; (3) objectives; and (4) constraints. Different input parameters are provided by network administrator or regulatory authorities. In CRNs, central controller generally knows about the SUs and relays in the network. In case of primary and secondary network cooperation, it is possible that CRNs central controller has the information about the number of PUs and their respective geographical locations. Interference threshold is set by the regulatory authorities. Value of interference threshold depends on the spectrum sharing regime. Knowledge of channel state information (CSI) is a significant input parameter. Most of the RA algorithms assume that CSI is known at both transmitter and receiver. Under different constraints, the RA schemes are analyzed and evaluated where performance is measured in different metrics in order to produce desired outputs.

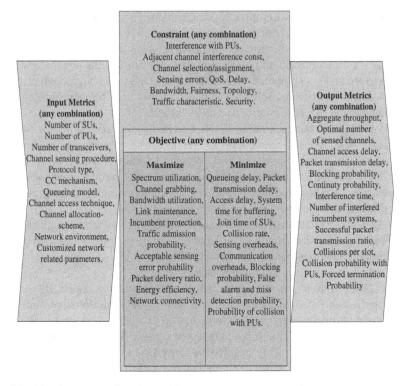

Fig. 3.2 Generic resource allocation problem

3.4 Resource Allocation Strategies

In this section, we explore and review the optimization strategies for RA in the context of CRNs. Depending upon the objectives, these strategies can be categorised under several performance optimization criteria. In the following, we provide a comprehensive overview of these categories and optimization strategies, and highlight the advantages, disadvantages, and the application areas associated to each of them.

3.4.1 Heuristic

In some cases it is tremendously tough, or even not possible, to get the optimal solution of certain problems. Hence, one may impose non-realistic assumptions to simplify the problem structure, and then find the optimal solution for the simplified structure. However, it is rather better to look for a reasonable (non-optimal) solution for the real problem structure than searching the optimal solution for inaccurate oversimplified structure. Heuristic strategies give logical solutions, that are not guaranteed to be the optimal, with acceptable time and space complexity [21]. Additionally, heuristic strategies are less restrictive than optimization methods, and thus they allow the usage of structures that are more presentable to the real world problems [22]. For CRNs, optimal solutions for RA problems are typically of high complexity. Hence, heuristic strategies are appropriate to find a fine solution quickly. Usually, there is no methodical structure for developing heuristic strategies. Thus, researchers build up their own methods according to the problems found in the literature [21].

Several efforts (e.g. [23–26]) proposed heuristic strategies for the RA problem in CRNs. The authors in [23] proposed two heuristic distributed RA algorithms based on opportunistic splitting. A heuristic packet transmission scheduling technique was developed in [24] that depends on measured link and interference conditions at CR nodes. In [25], a low complexity heuristic algorithm is proposed that efficiently solves the problem of spectrum sharing and downlink user scheduling. A two-step heuristic algorithm for sub-channel allocation and power distribution was developed in [26]. This algorithm extensively decreases the problem complexity while presenting an excellent approximate solution near optimal.

It is meaningful to mention that one disadvantage of heuristic strategies is that they are problem-specific and could not be universal. Also, there is no guarantee for convergence, especially for iterative strategies that may trapped into local optimal points far from the global optimal solution [18]. However, they stay on an essential substitution that can offer reasonable, fast, and yet easy solutions to be employed in many conditions. Additionally, heuristic solutions are more robust to problem variations than optimal solutions.

3.4.2 Optimization

Optimization problems intend to search the optimal of all possible solutions, which is the one that minimizes or maximizes a certain objective function. This objective function is a function used to assess a quality of the generated solution. The general mathematical structure of an optimization problem can be written as [27]:

$$\text{minimize} \quad f_0(x) \tag{3.1}$$
$$\text{subject to:} \quad f_i(x) \le b_i, \quad i = 1, \dots, m$$

where the problem components are as follows:

- $x = (x_1, \dots, x_n)$: optimization variables
- $f_0 : \mathbb{R}^n \to \mathbb{R}$: objective function
- $f_i : \mathbb{R}^n \to \mathbb{R}, i = 1, \dots, m$: constraint functions
- and the optimal solution x^* has the smallest value of f_0 among all vectors that satisfy the constraints

There are numerous types of objective functions occurring in RA for CRN of which three emerge to be very famous. Most of the other types are related to those three. The three are:

- Maximization of network sum throughput,
- Maximization of overall user data rate,
- Maximization of user capacity.

The other types of objective functions that have been studied in the literature include; maximization of spectrum utilizations, maximization of total channel utility, maximization of fairness scaling factor, maximization of downlink channel capacity, minimization of total transmission power, and minimization of total power consumption. Even the three stated earlier appear to have been utilized interchangeably by authors. Thus, all those three objective functions are combined together into a class of "capacity" measure. As for constraints, there are three that are usually occurring and they are:

- Power constraint,
- Constraint on interference,
- Minimum data rate for users.

The most commonly occurring decision variables are:

- Transmission power,
- Sub-channel allocation, which results in a zero-one decision variable.

In CRNs, optimization problems are generally used to formulate the RA problem for efficient utilization of the available resources. The formulated problems could be convex, non-convex, linear, or non-linear with different objectives such as sum-rate maximization, power minimization, fairness, etc. Certainly, complexity depends on the optimization problem model. For instance, convex optimization problems can be solved using standard optimization methods. Furthermore, a linear program (LP), wherein the objective function and constraints are all linear, is generally solved through iterative optimization problem solvers with less complexity and higher speed as compared to non-linear convex optimization problems with similar dimensions [27]. In non-linear programming (NLP), the objective function or some of the constraints are non-linear. For computationally intensive problems, greedy, heuristic algorithms appear as essential alternatives to accelerate the process of obtaining an acceptable approximate solution. For example, incorporating integer constraints in the problem such as the number of allocated time slots, makes it very complicated [28]. Generally, integer and mixed integer non-linear programming (MINLP) are computationally extensive (NP-hard) [29]. Hence, heuristic alternatives are usually utilized.

3.4.3   Graph Theory

Graph theory is one of the most extensively used tools for modeling and analyzing the interaction (or contention) in networks. A graph G consists of a set of vertices V and edges E, and is denoted by $G = (V, E)$. These components are mapped to the network elements according to the studied problem. Usually, vertices correspond to entities and edges represent the interaction between these entities. The graph theory could be employed to vast range of networks such as mechanical, transportation, and communication networks, among others. It can offer appropriate tools for solving network-related problems. Generally, graph theory is utilized when the network structure is known a priori [30].

In wireless communications, graph theory is largely used to solve the scheduling and RA problems, especially for problems of high computational effort [31]. In literature, the RA problem was solved using different types of graphs such as vertex-coloring graph, conflict graph, and bipartite graph [32, 33]. For CRNs, the SUs are generally mapped to the vertices and the edges mapping varies according the model definition to characterize certain relation between two vertices. For example, in a conflict or vertex-coloring graph model, an edge between two vertices (SUs' links) shows that the SUs are in the interference range of each other [25, 31]. Also, the RA for coloring graph problem is comparable to assigning each vertex a color (i.e., assign each link or SU).

If the vertex set V of a graph G can be split into two disjoint subsets $V1$, $V2$ such that each edge connects two vertices in different subsets, then G is a bipartite graph [34]. A bipartite matching problem is largely used for spectrum assignment and RA in CRNs [35–38]. Generally the two partitions of the bipartite graph are mapped to the SUs and the frequency bands available for assignment. The edges connecting two vertices (SU link and frequency band) for such a problem means that the SU requests (or accepts to be assigned) the corresponding frequency band.

In order to be application specific, coloring and conflict graphs are more appropriate for interference-limited environment. This is similar to the cases of having multiple SUs transmit simultaneously, and thus they can generate high interference to each other. Therefore, using the prescribed conflict or coloring graphs, also called node contention graphs (NCG), assists in modeling this interference in a way that makes simpler the development of a proper RA [31, 39]. On the other hand, the works adopted bipartite graphs usually allowed one SU per band with the owner PU to limit the interference in the system [35, 36]. However, adding preferences to the SUs and PUs guarantees better protection for the primary network as well as improved secondary performance. This is because the PUs' preferences are typically set according to their tolerated interference. Certainly, graph-based algorithms cannot integrate multiple performance metrics and so, normally, QoS is not guaranteed.

3.4.4 Game Theory

Game theory is a mathematical framework utilized to model and analyze the interaction among multiple individuals whose actions affect each other, or possibly who have conflict of interest [40, 41]. Usually, the game is modeled by considering a set of players (decision makers) who obtain the decisions (or actions) to maximize their payoff (utility). Although it is originally developed for economics, game theory found success in multiple fields such as engineering, biology, political science, physiology, etc. [42]. This is due to its advanced theoretical foundations which make it a powerful tool in decision making. Mathematically, the game is denoted as $G(\mathcal{N}, \mathcal{A}, \{u_i\})$, and the three game components are [42, 43]:

- $\mathcal{N} = \{1, 2, \ldots, N\}$ represents the finite set of players,
- $\mathcal{A} = A_1 \times A_2 \times \ldots A_N$ represents the sets of actions such that A_i is the set of actions available for player i,
- $u_i : A \rightarrow \mathbb{R}$ represents payoff/utility function of player i, which is a function of actions of all players.

Generally, game theory can be classified into two broad schemes: non-cooperative and cooperative schemes. These schemes have been broadly studied in literature. In a non-cooperative scheme, the players act in a selfish manner to maximize their own utility regardless of the impact of this on others. Nash Equilibrium (NE) is commonly known to be the solution point of non-cooperative game. In a cooperative game model, collaboration among the players and information exchanging are the ways to enhance the overall utility of the network and maximize their mutual benefits. Nash Bargaining (NB) is the common solution point for this game model. Also, other game categories exist such as static/dynamic, and complete/incomplete information.

In wireless communications, game theory is utilized to model and analyze several functions related to different layers (physical, data link, network, etc.). Usually, game theoretic approaches are used for power control, interference management, resource allocation, network selection in an environment of multiple deployed technologies, hand-off management, among others. These functions can be employed in the context of wireless sensor networks, CRNs, ad-hoc networks, etc. One major advantage of using non-cooperative game theory in wireless communication is its ability to offer robust and efficient distributed algorithms that rely only on local information and reduce the signaling overhead and computation complexity of the centralized approaches. However, the game model should be formulated carefully to be able to search a stable solution.

Unlike users of conventional wireless networks, CR users are intelligent and have the ability to observe, learn, interact with the surrounding environment, and adjust their parameters accordingly for an optimized performance. However, these users can be belonging to different authorities or have different goals. Thus, cooperation between them cannot be taken for granted. Instead, these users may act selfishly for each one to maximize its own payoff. Generally, an underlay CR spectrum sharing game model includes the following:

- Players: represented by the set of SUs or both SUs and PUs,
- Actions or strategies: depend on the player whether it is an SU or a PU,
- Utilities: an objective function to be maximized such as throughput, SINR, QoS, fairness, revenue, etc.

Thus, game theory occurs as a very important tool that can efficiently model and analyze the spectrum sharing process among the users that have conflict of interests in CRNs. Furthermore, game theory helps to design self-organized algorithms that are more appropriate to the dynamic wireless nature.

3.4.5 Genetic Algorithms

Genetic Algorithm is adaptive search algorithm based on the evolutionary ideas of natural selection [44]. An iterative process starts with a randomly generated set of solutions called population. Best individuals are selected through the utility function (called here fitness function). Then, starting from this subset, a second population is produced through genetic operators: crossover and/or mutation. The new population shares many of the characteristics of its parents, and it hopefully represents a better solution. The algorithm typically terminates when it converges to the optimal solution or after a fixed number of iterations. Genetic algorithms are chosen to solve RA problems due to their fast convergence and the possibility of obtaining multiple solutions. There are quite a few efforts [45–47] are found in the literature that utilizes genetic algorithms to solve RA problems.

3.4.6 Particle Swarm Intelligence Algorithms

Particle swarm intelligence (PSI) algorithms are population based stochastic optimization technique inspired by the collective behaviour of social biological individuals (e.g. bird flocking or fish schooling). PSI algorithms model network users as a population of simple agents interacting with the surrounding environment. Each individual has relatively little intelligence, however, the collaborative behaviour of the population directs to a global intelligence, which allows to solve complex tasks. For example, in social insect colonies, different activities are often executed by those individuals that are better equipped for the task. This phenomenon is named division of labour [48]. PSI algorithms are scalable, fault tolerant and moreover, they adapt to changes in real time. There are quite a few efforts [49–52] are found in the literature that utilizes PSI algorithms to solve RA problems.

3.4.7 Geometric Water-Filling Method

The conventional water-filling (CWF) [53, 54] problem can be described as follows. Given $P_T > 0$, as the total signal power (or volume of the water); the allocated power and the propagation path gain for the ith channel are given as P_i and h_i respectively where, $i = 1 \ldots N$; and N is the total number of subcarriers. Now without loss of generality, if $\{h_i\}_{i=1}^N$ be positive and monotonically decreasing, then the optimization problem can be written as:

$$\max_{\{P_i\}} \sum_{i=1}^N \log(1 + h_i P_i) \qquad (3.2)$$

subject to: $C1 : \ 0 \le P_i, \forall i;$

$$C2 : \ \sum_{i=1}^N P_i = P_T$$

where, constraint C1 is the allocated power to be nonnegative and constraint C2 is the total power constraint. To find the solution to problem (3.2), we usually start from the Karush-Kuhn-Tucker (KKT) conditions of the problem, as a group of the optimality conditions. The water level (μ) needs to be chosen to satisfy the power sum constraints with equality ($\sum_{i=1}^N P_i = P_T$) to find the optimal solution.

In [55], GWF approach is proposed to solve the CWF problem and its weighted form. It has two advantages, they are: (1) the geometric approach can compute the exact solution to the CWF, including the weighted case, with less computation and easier analysis without determining the water level through solving the non-linear system, (2) machinery of the proposed geometric approach can overcome the limitations of the CWF algorithm to include more stringent constraints.

Figure 3.3 gives an illustration of the GWF algorithm proposed in [55]. Suppose there are 4 steps/stairs ($N = 4$) with unit width inside a water tank. For the conventional approach, the dashed horizontal line, which is the water level μ, needs to be determined first and then the power allocated (water volume) above is solved. Let λ_i denotes the "step depth" of the ith stair which is the height of the ith step to the bottom of the tank, and is given as

$$\lambda_i = \frac{1}{h_i} \ \text{ for } i = 1, 2, \ldots, N. \qquad (3.3)$$

Since the sequence h_i is sorted as monotonically decreasing, the step depth of the stairs indexed as $[1, \ldots, N]$ is monotonically increasing. Now $\delta_{i,j}$ be the "step depth difference" of the ith and the jth stairs, can be expressed as,

$$\delta_{i,j} = \lambda_i - \lambda_j = \frac{1}{h_i} - \frac{1}{h_j} \ \text{ as } i \ge j \text{ and } 1 \le i, j \le N. \qquad (3.4)$$

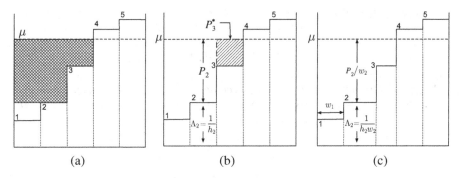

Fig. 3.3 Illustration for the proposed GWF algorithm (**a**) Illustration of $P_t(n)$ (shadowed area, representing the total water/power above step n) when $n = 2$, (**b**) Illustration of water level step $n^* = 3$, allocated power for the third step P_3^*, and step/stair depth $\Lambda_i = \frac{1}{h_i}$ and (**c**) Illustration of the weighted case

Instead of trying to determine the water level μ, which is a real nonnegative number, the water level step, denoted by n^*, is the target to solve, that is the highest step under water. Based on the result of n^*, the solutions for power allocation, can be written instantly. Let $P_t(n)$ denote the water volume above step n or zero, whichever is greater. The value of $P_t(n)$ can be solved by subtracting the volume of the water under step n from the total power P_T, as,

$$
\begin{aligned}
P_t(n) &= \left\{ P_T - \left[\sum_{i=1}^{n-1} \left(\frac{1}{h_n} - \frac{1}{h_i} \right) \right] \right\}^+ \\
&= \left\{ P_T - \left[\sum_{i=1}^{n-1} \delta_{n,i} \right] \right\}^+, \quad \text{for } n = 1, \dots, N
\end{aligned}
\tag{3.5}
$$

Due to the definition of $P_t(n)$ being the power (water volume) above step n, it cannot be a negative number. Therefore we use $\{.\}^+$ in (3.5) to assign 0 to $P_t(n)$ if the result inside the bracket is negative. The corresponding geometric meaning is that the nth level is above water.

According to [55], the explicit solution to (3.2) is:

$$
P_i = \begin{cases} P_{n^*} + (\lambda_{n^*} - \lambda_i) & 1 \le i \le n^* \\ 0, & n^* < i \le N, \end{cases}
\tag{3.6}
$$

where the water level step n^* is given as

$$
n^* = \max\left\{ n \,|\, P_t(n) > 0, 1 \le n \le N \right\}
\tag{3.7}
$$

and the power level for this step is

$$P_{n^*} = \frac{1}{n^*} P_t(n^*)$$ (3.8)

1. *Weighted Geometric Water-Filling Method (WGWF)*: GWF method for the weighted case, is also proposed in [55]. If weighted coefficients $w_i > 0$, $\forall i$ and without loss of generality, $\{h_i.w_i\}_{i=1}^{N}$ being positive and monotonically decreasing, then the optimization problem can be

$$\max_{\{P_i\}} \sum_{i=1}^{N} w_i \log(1 + h_i P_i)$$ (3.9)

subject to: $C1: 0 \le P_i, \forall i$;

$$C2: \sum_{i=1}^{N} P_i = P_T$$

where, constraint C1 is the allocated power to be nonnegative and constraint C2 is the total power constraint.

Let, the width of the ith step is denoted as w_i and the value of $\frac{1}{h_i}$ denotes the volume under the ith step to the bottom of the tank then, the step depth of the ith step is given as,

$$\lambda_i = \frac{1}{h_i w_i} \text{ for } i = 1, 2, \ldots, N.$$ (3.10)

Then, $P_t(n)$, the water volume above step n, can be obtained using the similar approach as in the previous section considering the step depth difference and the width of the stairs as,

$$P_t(n) = \left\{ P_T - \left[\sum_{i=1}^{n-1} \left(\frac{1}{h_n} - \frac{1}{h_i} \right) w_i \right] \right\}^{+}$$

$$= \left\{ P_T - \left[\sum_{i=1}^{n-1} \delta_{n,i} \right] w_i \right\}^{+}, \text{ for } n = 1, \ldots, N$$ (3.11)

According to [55], the explicit solution to (3.9) is:

$$P_i = \begin{cases} [\frac{P_{n^*}}{w_{n^*}} + (\lambda_{n^*} - \lambda_i)]w_i & 1 \le i \le n^* \\ 0, & n^* < i \le N, \end{cases}$$ (3.12)

where the water level step n^* is given as

$$n^* = \max\left\{ n \mid P_t(n) > 0, 1 \le n \le N \right\} \tag{3.13}$$

and the power level for this step is

$$P_{n^*} = \frac{w_{n^*}}{\sum_{i=1}^{n^*} w_i} P_t(n^*) \tag{3.14}$$

2. *Weighted Geometric Water-Filling with Individual Peak Power Constraints (GWFPP)*: The weighted GWFPP problem is stated in [55] as follows. Given $P_T > 0$, as the total signal power (or volume of the water); the allocated power and the propagation path gain for the ith channel are given as P_i and h_i respectively where, $i = 1 \ldots N$; and N is the total number of subcarriers. Also, the weights $w_i > 0$, $\forall i$ and without loss of generality, $\{h_i.w_i\}_{i=1}^{N}$ being positive and monotonically decreasing, then the optimization problem can be

$$\max_{\{P_i\}} \sum_{i=1}^{N} w_i \log(1 + h_i P_i) \tag{3.15}$$

subject to: $C1 : 0 \le P_i \le S_i, \forall i;$

$$C2 : \sum_{i=1}^{N} P_i \le P_T$$

where, the constraint C1 of $0 \le P_i$ in (3.9) is extended to $0 \le P_i \le S_i$, i.e., additional individual peak power constraints, and $\sum_{i=1}^{N} P_i = P_T$ to $\sum_{i=1}^{N} P_i \le P_T$.

The problem (3.15) is thus referred to as (weighted) geometric water-filling with sum and individual peak power constraints (WGFPP).

The operation of GWFPP algorithm according to [55] is stated as follows:

Input: vector $\{\lambda_i\}, \{w_i\}, \{S_i\}$ for $i = 1, 2, \ldots, N$, the set $E = 1, 2, \ldots, N$ and P_T

1) Compute $\{P_i\}$ utilizing (3.12)–(3.14).
2) The set Λ is defined by the set $\{i \mid P_i > S_i, i \epsilon E\}$. If Λ is empty set, output $\{P_i\}_{i=1}^{N}$; else $P_i = S_i$, as $i \epsilon \Lambda$.
3) Update E with $E \setminus \Lambda$ and P_T with $P_T - \sum_{i \epsilon \Lambda} P_i$. Then return to 1) of the GWFPP.

References

1. W. Xiong, A. Mukherjee, H.M. Kwon, MIMO cognitive radio user selection with and without primary channel state information. IEEE Trans. Veh. Technol. **65**(2), 985–991 (2016)
2. E. Driouch, W. Ajib, Downlink scheduling and resource allocation for cognitive radio MIMO networks. IEEE Trans. Veh. Technol. **62**(8), 3875–3885 (2013)
3. S. Wang, F. Huang, C. Wang, Adaptive proportional fairness resource allocation for OFDM-based cognitive radio networks. Wirel. Netw. **19**(3), 273–284 (2013)
4. L. Lu, D. He, X. Yu, G.Y. Li, Energy-efficient resource allocation for cognitive radio networks, in *Proceedings of the IEEE Global Communications Conference (GLOBECOM)*, Atlanta, December 2013, pp. 1026–1031
5. H. Guo, V.C.M. Leung, Orthogonal transmissions for spectrum underlay MISO cognitive radio. IEEE Trans. Wirel. Commun. **11**(4), 1266–1270 (2012)
6. A.G. Marques, L.M. Lopez-Ramos, G.B. Giannakis, J. Ramos, Resource allocation for interweave and underlay CRs under probability-of-interference constraints. IEEE J. Sel. Areas Commun. **30**(10), 1922–1933 (2012)
7. L.B. Le, E. Hossain, Resource allocation for spectrum underlay in cognitive radio networks. IEEE Trans. Wirel. Commun. **7**(12), 5306–5315 (2008)
8. J.R. Gállego, M. Canales, J. Ortín, Distributed resource allocation in cognitive radio networks with a game learning approach to improve aggregate system capacity. Elsevier Ad Hoc Netw. **10**(6), 1076–1089 (2012)
9. Y. Xu, X. Zhao, Distributed power control for multiuser cognitive radio networks with quality of service and interference temperature constraints. Wirel. Commun. Mob. Comput. **15**(14), 1773–1783 (2015)
10. S. Gong, P. Wang, L. Duan, Distributed power control with robust protection for PUs in cognitive radio networks. IEEE Trans. Wirel. Commun. **14**(6), 3247–3258 (2015)
11. M. Rasti, M. Hasan, L.B. Le, E. Hossain, Distributed uplink power control for multi-cell cognitive radio networks. IEEE Trans. Commun. **63**(3), 628–642 (2015)
12. J. Zhang, Z. Zhang, H. Luo, A. Huang, R. Yin, Uplink scheduling for cognitive radio cellular network with primary user's QoS protection, in *Proceedings of the IEEE Wireless Communications and Networking Conference (WCNC)*, Sydney, April 2010, pp. 1–5
13. R. Sumathi, M. Poornima, M. Suganthi, User aware mobility management in cognitive radio cellular network, in *Proceedings of the International Conference on Electronics and Communication Systems (ICECS)*, Coimbatore, February 2014, pp. 1–5
14. C.R. Stevenson, G. Chouinard, Z. Lei, W. Hu, S.J. Shellhammer, W. Caldwell, IEEE 802.22: the first cognitive radio wireless regional area network standard. IEEE Commun. Mag. **47**(1), 130–138 (2009)
15. A. Dimogiorgi, W. Hamouda, A proposed enhanced scheme for the dynamic frequency hopping performance in the IEEE 802.22 standard. Wirel. Commun. Mob. Comput. J. **16**(16), 2714–2729 (2016)
16. M.E. Tanab, W. Hamouda, Y. Fahmy, On the distributed resource allocation of MIMO cognitive radio networks, in *Proceedings of the IEEE Global Communications Conference (GLOBECOM)*, San Diego, December 2015, pp. 1–6
17. R. Jurdak, C.V. Lopes, P. Baldi, A survey, classification and comparative analysis of medium access control protocols for ad hoc networks. IEEE Commun. Surv. Tutorials **6**(1), 2–16 (2004)
18. I.F. Akyildiz, W.-Y. Lee, K.R. Chowdhury, CRAHNs: cognitive radio ad hoc networks. Ad Hoc Netw. **7**(5), 810–836 (2009)
19. S.-J. Kim, G.B. Giannakis, Optimal resource allocation for MIMO ad hoc cognitive radio networks. IEEE Trans. Inf. Theory **57**(5), 3117–3131 (2011)
20. I.F. Akyildiz, W.-Y. Lee, K.R. Chowdhury, Spectrum management in cognitive radio ad hoc networks. IEEE Netw. **23**(4), 6–12 (2009)
21. N. Kokash, An introduction to heuristic algorithms. Department of Informatics and Telecommunications (2005), pp. 1–8

22. E.A. Silver, An overview of heuristic solution methods. J. Oper. Res. Soc. **55**, 936–956 (2017)
23. M.E. Tanab, Y. Fahmy, M.M. Khairy, Opportunistic splitting algorithm for underlay cognitive radio networks, in *Proceedings of the IEEE Sixth International Conference on Ubiquitous and Future Networks (ICUFN)*, July 2014, pp. 9–14
24. B. Wang, D. Zhao, J. Cai, Joint connection admission control and packet scheduling in a cognitive radio network with spectrum underlay. IEEE Trans. Wirel. Commun. **10**, 3852–3863 (2011)
25. E. Driouch, W. Ajib, Downlink scheduling and resource allocation for cognitive radio MIMO networks. IEEE Trans. Veh. Technol. **62**, 3875–3885 (2013)
26. S. Wang, F. Huang, C. Wang, Adaptive proportional fairness resource allocation for OFDM-based cognitive radio networks. Wirel. Netw. **19**, 273–284 (2013)
27. S. Boyd, L. Vandenberghe, *Convex Optimization* (Cambridge University Press, Cambridge, 2004)
28. L. Fu, M. Johansson, M. Bengtsson, Energy efficient transmissions in cognitive MIMO systems with multiple data streams. IEEE Trans. Veh. Technol. **14**, 5171–5184 (2015)
29. R. Kannan, C.L. Monma, *On the Computational Complexity of Integer Programming Problems* (Springer, Berlin, 1978)
30. E.Z. Tragos, S. Zeadally, A.G. Fragkiadakis, V.A. Siris, Spectrum assignment in cognitive radio networks: a comprehensive survey. IEEE Commun. Surv. Tutorials **15**(3), 1108–1135 (2013)
31. E. Driouch, W. Ajib, A.B. Dhaou, A greedy spectrum sharing algorithm for cognitive radio networks, in *Proceedings of the IEEE International Conference on Computing, Networking and Communications (ICNC)*, January 2012, pp. 1010–1014
32. Y.-J. Chang, Z. Tao, J. Zhang, C.-C.J. Kuo, A graph-based approach to multi-cell OFDMA downlink resource allocation, in *Proceedings of the IEEE Global Telecommunications Conference (GLOBECOM)*, December 2008, pp. 1–6
33. Y. Liu, M. Tao, Optimal channel and relay assignment in OFDM-based multi-relay multi-pair two-way communication networks. IEEE Trans. Commun. **60**, 317–321 (2012)
34. R.J. Wilson, *Introduction to Graph Theory*, 4th edn. (Prentice Hall, Harlow, 1996)
35. L. Lu, D. He, X. Yu, G.Y. Li, Energy-efficient resource allocation for cognitive radio networks, in *Proceedings of the IEEE Global Telecommunications Conference (GLOBECOM)*, December 2013, pp. 1026–1031
36. L. Lu, D. He, X. Yu, G.Y. Li, Graph-based robust resource allocation for cognitive radio networks, in *Proceedings of the IEEE International Conference on Acoustics, Speech and Signal Processing (ICASSP)*, May 2014, pp. 7298–7302
37. W. Zhou, T. Jing, W. Cheng, T. Chen, Y. Huo, Combinatorial auction based channel allocation in cognitive radio networks, in *Proceedings of the IEEE International Conference on Cognitive Radio Oriented Wireless Networks (CROWNCOM)*, July 2013, pp. 135–140
38. L. Chen, L. Huang, H. Xu, H. Deng, Z. Sun, Optimal channel assignment schemes in underlay CRNs with multi-PU and multi-SU transmission pairs. Wirel. Algorithms Syst. Appl. **9204**, 29–39 (2015)
39. C. Peng, H. Zheng, B.Y. Zhao, Utilization and fairness in spectrum assignment for opportunistic spectrum access. Mob. Netw. Appl. **11**, 555–576 (2006)
40. K. Akkarajitsakul, E. Hossain, D. Niyato, D.I. Kim, Game theoretic approaches for multiple access in wireless networks: a survey. IEEE Commun. Surv. Tutorials **13**(3), 372–395 (2011)
41. M. Felegyhazi, J.-P. Hubaux, Game theory in wireless networks: a tutorial. IEEE Communications Surveys and Tutorials, No. LCA-REPORT-2006-002 (2006)
42. B. Wang, Y. Wu, K.R. Liu, Game theory for cognitive radio networks: an overview. Elsevier Comput. Netw. **54**, 2537–2561 (2010)
43. O.B. Abdulghfoor, M. Ismail, R. Nordin, Application of game theory to underlay ad-hoc cognitive radio networks: an overview, in *Proceedings of the IEEE International Conference on Space Science and Communication (IconSpace)* 296–301 (2013)
44. A.D. Domenico, E.C. Strinati, M.D. Benedetto, A survey on MAC strategies for cognitive radio networks. IEEE Commun. Surv. Tutorials **14**(1), 21–44 (2012)

45. H.-S. Lang, S.-C. Lin, W.-H. Fang, Subcarrier pairing and power allocation with interference management in cognitive relay networks based on genetic algorithms. IEEE Trans. Veh. Technol. **65**, 7051–7063 (2016)
46. Y. Jiao, I. Joe, Energy-efficient resource allocation for heterogeneous cognitive radio network based on two-tier crossover genetic algorithm. J. Commun. Netw. **18**(1), 112–122 (2016)
47. Y.E. Morabit, F. Mrabti, E.H. Abarkan, Spectrum allocation using genetic algorithm in cognitive radio networks, in *Third International Workshop on RFID and Adaptive Wireless Sensor Networks (RAWSN)* 90–93 (2015)
48. B. Atakan, O.B. Akan, Biologically-inspired spectrum sharing in cognitive radio networks, in *Proceedings of the IEEE Wireless Communications and Networking Conference (WCNC)*, March 2007, pp. 43–48
49. S. Yang, J. Wang, Y. Han, Q. Zhao, Dynamic spectrum allocation algorithm based on fairness for smart grid communication networks, in *Proceedings of the IEEE International Conference of Chinese Control Conference (CCC)*, July 2016, pp. 6873–6877
50. S.B. Behera, D.D. Seth, Resource allocation for cognitive radio network using particle swarm optimization, in *Proceedings of the IEEE International Conference on Electronics and Communication Systems (ICECS)*, February 2015, pp. 665–667
51. B. Chen, M. Zhao, M. Lei, L. Zhang, An optimal spectrum sharing method for MIMO cognitive radio networks, in *Proceedings of the IEEE International Symposium on Personal, Indoor, and Mobile Radio Communications (PIMRC)*, September 2015, pp. 1007–1011
52. P. Tiwari, S. Saha, Co-channel interference constrained spectrum allocation with simultaneous power and network capacity optimization using PSO in cognitive radio network, in *Proceedings of the IEEE International Conference on Advanced Networks and Telecommunications Systems (ANTS)*, December 2015, pp. 1–3
53. D.P. Palomar, J.R. Fonollosa, Practical algorithms for a family of water-filling solutions. IEEE Trans. Signal Process. **53**, 686–695 (2005)
54. W. Yu, J.M. Cioffi, On constant power water-filling, in *Proceedings of the IEEE International Conference on Communications*, vol. 6, June 2001, pp. 1665–1669
55. P. He, L. Zhao, S. Zhou, Z. Niu, Water-filling: a geometric approach and its application to solve generalized radio resource allocation problems. IEEE Trans. Wirel. Commun. **12**(7), 3637–3646 (2013)

Chapter 4
Resource Allocation in OFDM-Based Cognitive Radio Systems

4.1 Literature Review

Resource allocation problem in OFDM based CRNs has been widely studied under different settings in the open literature. A power allocation grouping scheme based on the interference channel gain, pulse shape and frequency distance is presented in [1] in order to improve capacity while the interference power for PUs stays at constant level. At the first stage, power is assigned to some groups based on the grouping scheme and at the second stage, the remaining power is allocated to others with water-filling algorithm.

In order to maximize the SUs' transmission capacity by adjusting the power allocation across the subcarriers, two suboptimal schemes for subcarrier allocation with associated triangular and Gaussian power loading mechanism are studied in [2]. Here, these schemes maintain the total power used within the power budget and also the interference at the PUs within a permissible threshold limit.

A low-complexity algorithm using power-increment or power-decrement WF processes is proposed in [3]. This algorithm is based on constraining the total and transmit powers of each subchannel.

In order to maximize the download capacity of CRNs, a subcarrier and power allocation algorithm based on the linear water-filling (LWF) scheme is proposed in [4] while keeping both the total transmit power and interference introduced to the PU below the pre-constraints.

A suboptimal power allocation algorithm for multiuser OFDM-based CRNs is presented in [5] that simplifies the procedure of determining the water level to allocate the power for each subcarrier under a peak power constraint. This method not only reduces the computational complexity but also can easily be combined with a subcarrier allocation algorithm for joint subcarrier and power allocation.

X. Fernando et al., *Cooperative Spectrum Sensing and Resource Allocation Strategies in Cognitive Radio Networks*, SpringerBriefs in Electrical and Computer Engineering, https://doi.org/10.1007/978-3-319-73957-1_4

Another suboptimal power allocation algorithm for multiuser OFDM-based CRNs is presented in [6] that simplifies the procedure of determining the water level to allocate the power for each subcarrier under a peak power constraint. This method not only reduces the computational complexity but also can easily be combined with a subcarrier allocation algorithm for joint subcarrier and power allocation.

A low-complexity suboptimal power allocation algorithm is proposed in [7] where CR pairs (CRPs) may use both nonactive and active PU pair (PUP) bands as long as the total co-channel interference (CCI) and cross-channel interference (XCI) do not exceed prescribed limits. The complexity reduction is done by making a validated approximation based on the following: (1) the fact that XCI from CRUs to PUPs is mainly limited to a few subchannels adjacent to the PUP bands and (2) the assumption that the bandwidth for a PUP is typically much larger than that of an OFDM subchannel and that there is usually a guard band between two adjacent PUP bands.

An optimal scheme employing joint power allocation in overlay and underlay fashion has been proposed in [8]. The optimal scheme based on the Lagrange formulation maximizes the downlink capacity of CR users while maintaining a total power budget and keeping the interference introduced to the PU band below a threshold. A low complexity suboptimal scheme is also presented where equal low power is allocated to all the underlay subcarriers and a ladder profile is used to load power in the overlay subcarriers.

For weighted sum rate maximization in CRNs, a power limited multilevel WF iterative algorithm has been proposed in [9] where subcarrier power constraints were added to the traditional problem.

With the total power constraint and the power constraint on each subchannel, an iterative partitioned water-filling (IPWF) algorithm was proposed in [10] to realize optimal power allocation in OFDM based CRNs.

Table 4.1 provides different approaches for power allocation in OFDM-based CRNs

4.2   Cognitive Radio System Model and Transmission Power Constraint

A typical cognitive radio systems is shown in Fig. 4.1 where PUs and SUs share the same bandwidth. In order to avoid harmful interference to each other, the SU needs to detect the opportunities when PUs are not utilizing the spectrum. Higher detection probability without errors provide successful exploitation of opportunities for transmission. In Fig. 4.1, a disk propagation model is considered to illustrate SU transmission. SU can detect any PU's activity within the detection region. However, those PUs that fall outside the detection region (like PU2 in Fig. 4.1), are undetectable by the SU. To deal with this situation, as in [11], PU2 defines a protection region with radius P and needs to maintain a certain interference

Table 4.1 Different approaches for power allocation in OFDM-based CRNs

Ref.	Objective	Constraints	Solution approach
[1]	Improve capacity while the interference power for PUs stays at constant level	Total transmit power and interference constraints	A grouping scheme based on the interference channel gain, pulse shape and frequency distance is considered
[2]	Maximize transmission capacity	Total transmit power and interference constraints	Two suboptimal schemes for subcarrier allocation with associated triangular and Gaussian power loading mechanism are studied
[3]	Maximize capacity	Total transmit power and individual subchannels' power constraints	A low-complexity algorithm using power-increment or power-decrement water-filling processes is considered
[4]	Maximize downlink capacity	Total transmit power and interference constraints	An iterative suboptimal power allocation algorithm based on LWF is proposed along with subcarrier allocation
[5]	Maximize overall throughput while keep the average interference to the PU within a target outage probability level	Total transmit power and interference constraints	WF process runs only once, and then directly calculates the final result without the need for searching the Lagrange multiplier
[6]	Maximize uplink capacity	Total transmit power and peak power constraints	A simple and low complexity suboptimal scheme is considered for determining the water level to allocate the power for each subcarrier
[7]	Maximize the overall rate while keeping the interference experienced by the PU pairs below certain thresholds	Total transmit power and interference constraints	A low-complexity suboptimal power allocation algorithm is proposed where CR pairs may use both nonactive and active PU pair bands as long as the total cochannel interference and cross-channel interference do not exceed prescribed limits
[8]	Maximize downlink capacity while keeping the interference introduced to the PU band below a threshold	Total transmit power and interference constraints	A low complexity suboptimal scheme for joint overlay and underlay power allocation is presented where equal low power is allocated to the underlay subcarriers and a ladder profile is used to load power in the overlay subcarriers
[9]	Maximize the weighted sum rate of SUs in the downlink, without causing interferences to PUs	Total transmit power and interference constraints	A power limited multilevel WF iterative algorithm is considered
[10]	Maximize the capacity maintaining the PU's interference limits	Total transmit power and per subchannel power constraints	An iterative partitioned water-filling algorithm is proposed

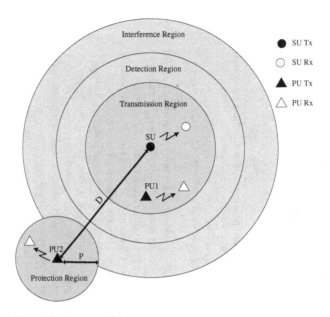

Fig. 4.1 Cognitive radio system model

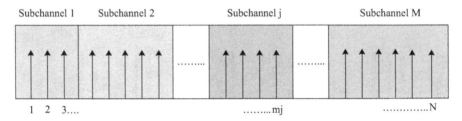

Fig. 4.2 Spectrum of SU in OFDM-based CRNs

level η within this area. In this scenario, SU's transmission power P_{tx}, subjected to interference constraint, can be written as [12]:

$$P_{tx} \leq \eta(D - P)^{\beta} \qquad (4.1)$$

where D is the distance between the SU transmitter and the nearest undetectable PU transmitter, β is the path attenuation factor and η is the maximum allowable interference level. OFDM is a potential technology in terms of modulation and power control. It is also highly flexible due to reconfigurable subcarrier structure to fit in CRNs for efficient utilization of spectrum opportunities. Figure 4.2 depicts a typical spectrum structure in OFDM-based CRNs. There are M subchannels licensed to M PU systems that can be potentially used by the SU based on opportunity detection. There are N subcarriers that are distributed among the M subchannels. For example, let the jth subchannel has total L_j subcarriers that can be utilized by the SU when PU is absent. For successful transmission, the SU first needs to test any PU transmitter in the desired subchannel. If found, the sum power of all the subcarriers in that subchannel will be set to zero until the PU transmission ends. If not, the SU can utilize this subchannel with the interference constraint described in (4.1). Let G_j is the interference constraint on the jth subchannel after spectrum detection, then,

$$G_j \triangleq \begin{cases} 0 & PU_j \text{ is detected} \\ \eta_j (D_j - P_j)^{\beta_j} & PU_j \text{ is not detected} \end{cases} \tag{4.2}$$

where η_j is the maximum allowable interference level for PU_j within the protection region whose radius is P_j, D_j is the distance between the SU's transmitter and the nearest undetectable PU_j's transmitter and β_j is the corresponding path attenuation factor.

4.3 Problem Formulation

Consider an OFDM communication system similar to [13]:

$$y_n[m] = h_n x_n[m] + w_n[m], \quad \text{where } n = 1, \ldots, N \tag{4.3}$$

where $x_n[m]$, $y_n[m]$, $w_n[m]$ and h_n are the input, output, noise signal and channel gain, respectively. Assume $P_T > 0$, as the total power constraint and P_i is the transmit power of an OFDM block, then the achievable data rate of reliable communication using the OFDM channel is,

$$R(P_i) = \sum_{i=1}^{N} B \log(1 + \frac{P_i |h_i|^2}{N_0}) \tag{4.4}$$

where N_0 is the noise power spectral density and B is the subcarrier spacing (bandwidth). Thus the power allocation has to be done to maximize the sum rate in (4.4). In OFDM based CRNs, the subchannel transmit power constraints impose further restrictions on the power allocation in addition to the total transmit power constraint. Assuming the allocated signal power and the propagation path gain for the ith subcarrier are given as P_i and h_i (where $\frac{|h_n|^2}{N_0} = h_i$) respectively, $i = 1, \ldots, N$ and, the weights $w_i > 0, \forall i$ then the optimal power allocation problem can be expressed as:

$$\max_{\{P_i\}} \sum_{i=1}^{N} w_i \log(1 + h_i P_i) \tag{4.5}$$

subject to: $C1: 0 \leq P_i \leq S_i, \forall i;$

$$C2: \sum_{i=1}^{N} P_i = P_T$$

$$C3: F_j \leq G_j, \forall j;$$

where, N is the total number of subcarriers and $j = 1, \ldots, M$; and M is the total number of subchannels, $F_j = \sum_{i=m_j}^{m_{j+1}-1} P_i, j \in A$ is the power allocated to the jth subchannel and m_j is the index of the first subcarrier and $m_{j+1} - 1$ is the index of the last subcarrier in the jth subchannel.

The power allocation problem in (4.5) is constructed considering three constraints. Constraint C1 consider individual peak power constraints, constraint C2 includes total power constraint and constraint C3 incorporates individual subchannel power constraints caused by the PUs's interference limits. Considering constraints C2 and C3, in [10] the authors proposed IPWF algorithm to obtain the optimal power allocation vector. Constraint C1 was introduced in [14] where the authors proposed GWFPP approach to solve the weighted radio resource allocation problems, that also provide optimal result. In this paper, we combined all those constraints to construct an optimal power allocation scheme.

4.4   Subcarrier Allocation

Assume I_m be the number of subcarriers allocated for one subchannel, is a variable. Since the utilization of any subchannel by the SU, is bounded by the power constraint on the jth subchannel (G_j), the transmit power for one subchannel on each subcarrier is given by

$$P_{m,i_m} = \frac{G_j}{I_m}, \quad i_m \in \hat{I}_m \tag{4.6}$$

where $\hat{I}_m$ represents the set of subcarriers allocated to one subchannel and I_m is the number of subcarriers allocated to one subchannel. The achievable data rate (using (4.4)) can be shown as

$$D = \sum_{m=1}^{M} \sum_{i_m \in I_m} R_{m,i_m}(P_{m,i_m}), \quad i_m \in \hat{I}_m \tag{4.7}$$

For a given set $\hat{I}_m$ for a subchannel, if one more subcarrier $i_m{}^*$ is allocated to that subchannel, i. e. $\hat{I}_m$ is replaced by $\hat{I}_m \cup \{i_m{}^*\}$, the change of achievable rate for that subchannel, $\Delta d_{m,i_m{}^*}(\hat{I}_m)$, can be given by

Fig. 4.3 Flow chart for subcarrier allocation

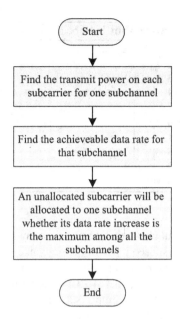

$$\Delta d_{m,i_m*}(\hat{I}_m) = D_m(\frac{G_j}{I_m + 1}) - D_m(\frac{G_j}{I_m})$$

$$= \sum_{i_m \in \hat{I}_m \cup \{i_m*\}} d_{m,i_m}(\frac{G_j}{I_m + 1}) - \sum_{i_m \in \hat{I}_m} d_{m,i_m}(\frac{G_j}{I_m}) \qquad (4.8)$$

In our proposed algorithm, assuming $\hat{I}_m$ has been allocated to one subchannel, whether an unallocated subcarrier i can be allocated to one subchannel, depends on whether its data rate increase $\Delta d_{m,i_m}(\hat{I}_m)$ is the maximum among all the subchannels. That is, subcarrier i will be allocated to $k^* = \max_{m,i_m} \Delta d_{m,i_m}(\hat{I}_m)$. The basics idea behind the subcarrier allocation is well depicted in Fig. 4.3 and the algorithm is described in Algorithm 1 description. After subcarrier allocation, in the second step, power allocation is carried out on its allocated subcarrier $\hat{A}_m$.

4.5 Power Allocation Using GWFPP

4.5.1 Iterative Partitioned Weighted Geometric Water-Filling with Individual Peak Power (IGPP)

In this work, utilizing IPWF [10] and GWFPP [14] approach, we proposed Iterative Partitioned Weighted Geometric Water-Filling with Individual Peak Power (IGPP) algorithm, that also provides optimal result. For IGPP the first step is to divide all the

---

Algorithm 1: Subcarrier allocation

---

1 initialization;
2 $\hat{I}_m = \phi,\, I_m = 0,\, \Delta d_{m,i_m}(I_m) = D_m(G_j)$
3 **for** $i \in N$ **do**
4   **for** $m \in M$ **do**
5   compute $\Delta d_{m,i_m}(\hat{I}_m)$ using (4.8)
6         $k^* = \max_{m,i_m} \Delta d_{m,i_m}(\hat{I}_m);$
7         $\hat{I}_m^* = \hat{I}_m \cup \{i\};$
8         $I_m^* = I_m^* + 1;$
9   **end**
10 **end**

---

subchannels into two sets, say A and B, and there are 2^M partitions in total. E set is a subsequence of the sequence $1, 2, \ldots, N$ and N is the total number of subcarriers. The next step would be to perform GWFPP for every jth subchannel and calculate P_i using (3.12)–(3.14). Let the set Λ is defined by the set $\{i \,|\, P_i > S_i, i \in E\}$. If Λ is an empty set, then P_i will be the output, otherwise, P_i will be set by the individual peak power S_i. Everytime we need to update the set E and the total power P_T. Since the finite set E is getting smaller and smaller until the set Λ is empty, algorithm GWFPP carries out at most, N loops to compute the optimal solution [14]. The next steps are to remove partitions that are represented by set D which is a set of subchannels that satisfies $F_j < G_j$ where $j \in A$. For each of the remaining partitions in the remainder set A, the geometric water-filling is performed on the subcarriers that belong to the jth subchannel. Algorithm IGPP carries out the loops to compute the optimal solution until the set A is empty. The last step would be to verify each partition whether they satisfy $F_j < G_j$ where $j \in B$. According to the paper [10], there is only one available partition and the corresponding power allocation vector is the solution. Based on the property of IGPP and the strict constraints on the objective function, there can be only one power allocation vector which satisfies all the conditions. In CWF problem, it is difficult to obtain the explicit expression of the optimal power allocation vector. A derived algorithm obtained from [10] and [14] forming IGPP, is described in Algorithm 2 description.

4.6 Performance Evaluation

The proposed algorithm IGPP along with the existing methods (IPWF, GWFPP), have been implemented using MATLAB. Randomized fading have been used to generate the outcomes of each algorithms. Here, we assumed that the distance between the SU and the PU is known, thus Eq. (4.2) has been used to calculate the subchannel power constraints (G_j). According to the subcarrier assignment strategy,

---

Algorithm 2: Iterative partitioned weighted geometric water-filling with individual peak power constraints (IGPP)

Input: vector $\{\frac{1}{h_i}\}$, $\{w_i\}$, $\{S_i\}$ for $i = 1, 2, \ldots, N$,
the set $E = 1, 2, \ldots, N$ and P_T

1 initialization;
2 $A = \{j | j = 1, 2, \ldots, M\}$
3 $B = \phi, F = \phi, \Lambda = \phi, P^* = P_T$;
4 $C = \{i | \text{the } i\text{th subcarrier belongs to the } j\text{th subchannel}, j \in A\}$;
5 **for** $j \in A$ **do**
6   **for** $i \in C$ **do**
7   compute $\{P_i\}$ using GWF
8          $\Lambda = \{i | P_i > S_i, i \in E\}$
9   **if** $\Lambda = \phi$ **then**
10   output: $\{P_i\}$;
11   **else**
12             $P_i = S_i, i \in A$;
13   **end**
14         $E = E \setminus \Lambda$
15   **end**
16     $F_j = \sum_{i=m_j}^{m_{j+1}-1} P_i, j \in A$
17     $D = \{j | F_j \leq G_j, j \in A\}$;
18     $L = \{i | \text{the } i\text{th subcarrier belongs to the } j\text{th subchannel}, j \in D\}$
19     $A = A \setminus D, B = B \cup D$
20     $C = C \setminus L, F = F \cup L$
21   **if** $A = \phi$ **then**
22   output: $\{P_i\}$;
23   **else**
24         $P^* = P^* - \sum_{i \in L} P_i$;
25   **end**
26 **end**

Table 4.2 Simulation parameters

Parameters	Values
No. of SU and PU	1, 4
Number of subchannels	4
Number of subcarriers	15
Protection region (P_j)	1 m
Path attenuation factor	2
Total power (P_T)	159
Bandwidth (B)	15 kHz
Subchannel power constraints (G_j)	{72, 44, 29, 17}

described in previous section, we got subcarriers 1, 2, 3, 4 to be in subchannel 1, subcarriers 5, 6, 7 to be in subchannel 2, subcarriers 8, 9, 10, 11 to be in subchannel 3 and the last set of subcarriers 12, 13, 14, 15 to be in subchannel 4. The simulation parameters are summarized in Table 4.2.

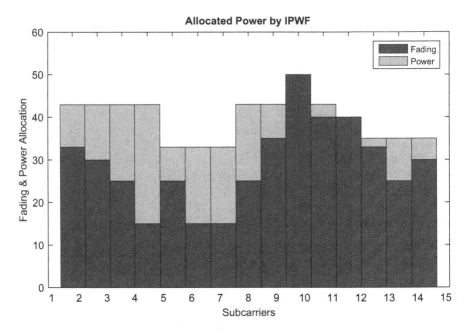

Fig. 4.4 Power allocation using iterative partitioned water-filling method

Figure 4.4 shows power allocation when IPWF method has been applied. Since IPWF mainly works based on partition of subchannels, the water levels are different for each subchannel due to the total subchannel power constraint for the partitions.

Figure 4.5 shows power allocation when GWFPP algorithm has been applied. This algorithm mainly consider individual peak power constraints. As a result, for each channel we get a different set of water level which follows the individual peak power constraints.

Figure 4.6 shows power allocation when IGPP algorithm has been applied. Due to the total subchannel power constraint and individual peak power constraints the simulation result varies from the results of IPWF and GWFPP. For each subcarrier we get a different set of water level which follows the individual peak power constraints but they are different from the GWFPP algorithm due to the total subchannel power constraints. Figure 4.7 shows power allocation by IGPP when considering different weight factors of subcarriers.

Figure 4.7 shows allocated power with corresponding noise power by IGPP when considering different weight factors for subcarriers. In X axis, cumulative summation of weighted coefficients are presented, as we moved towards right. In Fig. 4.7, the width of each bar varies due to the different weight factors of the subcarriers.

Figure 4.8 compares the allocated power for the IPWF, GWFPP and IGPP algorithms. For each subchannel, we sorted the subcarriers according to their noise power. Thus, the subcarrier that has highest noise power, comes first and

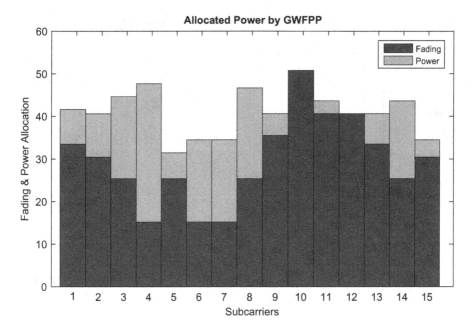

Fig. 4.5 Power allocation using geometric water-filling method with peak power constraints

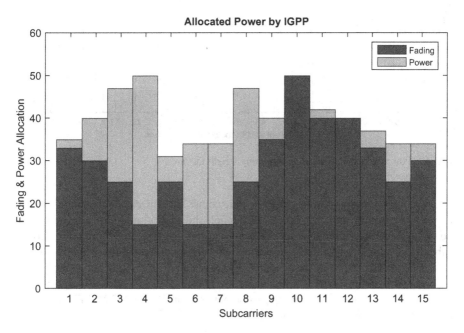

Fig. 4.6 Power allocation using iterative partitioned weighted geometric water-filling with individual peak power constraints algorithm

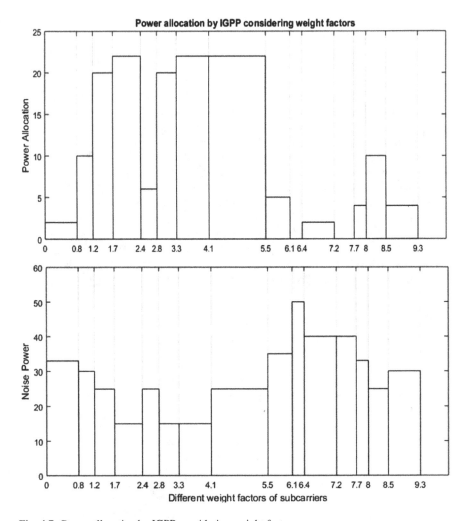

Fig. 4.7 Power allocation by IGPP considering weight factors

the subcarrier that has lowest noise power, comes last for each subchannel. From Fig. 4.8, we found that, for each subchannel, the algorithm IGPP allocated less power in the subcarrier that has highest noise power and allocated more power in the subcarrier that has lowest noise power than the other two algorithms. Thus, the IGPP allows better utilization of the available power resources than IPW and GWFPP.

Figure 4.9 compares the sum rates for the IPWF, GWFPP and IGPP algorithms. IGPP algorithm allows the sum rate to be better than IPW and GWFPP, due to the better utilisation of the power resources.

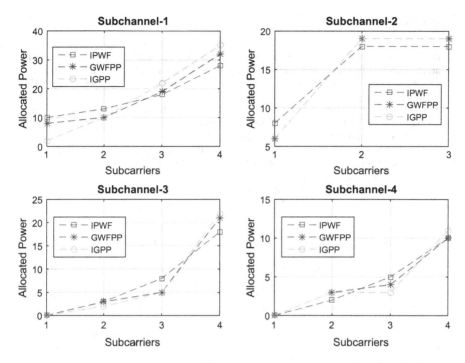

Fig. 4.8 Comparison of the sum rates for IPWF, GWFPP and IGPP

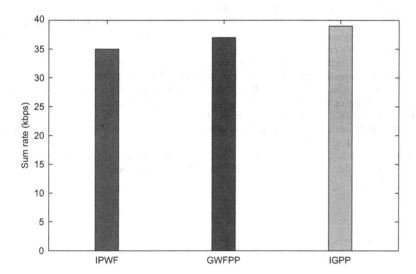

Fig. 4.9 Optimal power allocation vs subcarriers for each subchannel using IPWF, GWFPP and IGPP

References

1. E. Hosseini, A. Falahati, Power allocation grouping scheme considering constraints in two separate stages for OFDM-based cognitive radio system, in *Proceedings of the IEEE International Conference on Electrical Information and Communication Technology (EICT)*, February 2014, pp. 1–6
2. S. Chakraborty, P. Dhanuka, A. Kumar, S.P. Maity, Subcarrier and power allocation schemes for multiuser OFDM-based cognitive radio systems, in *Proceedings of the IEEE National Conference on Communications (NCC)*, February 2013, pp. 1–5
3. Q. Qi, A. Minturn, Y. Yang, An efficient water-filling algorithm for power allocation in OFDM-based cognitive radio systems, in *Proceedings of the IEEE International Conference on Systems and Informatics (ICSAI)*, May 2012, pp. 2069–2073
4. W. Jian, Y. Longxiang, L. Xu, Resource allocation based on linear waterfilling algorithm in CR systems, in *Proceedings of the IEEE Wireless Communications, Networking and Mobile Computing (WiCOM)*, September 2011, pp. 1–4
5. X. Zhou, B. Wu, P.-H. Ho, X. Ling, An efficient power allocation algorithm for OFDM based underlay cognitive radio networks, in *Proceedings of the IEEE Global Telecommunications Conference (GLOBECOM)*, December 2011, pp. 1–5
6. C.-H. Chen, C.-L. Wang, An efficient power allocation algorithm for multiuser OFDM-based cognitive radio systems, in *Proceedings of the IEEE Wireless Communications and Networking Conference (WCNC)*, April 2010, pp. 1–6
7. Y. Zhang, C. Leung, An efficient power-loading scheme for OFDM-based cognitive radio systems. IEEE Trans. Veh. Technol. **59**, 1858–1864 (2010)
8. G. Bansal, O. Duval, F. Gagnong, Joint overlay and underlay power allocation scheme for OFDM-based cognitive radio systems, in *Proceedings of the IEEE Wireless Communications and Networking Conference (WCNC)*, May 2010, pp. 1–5
9. Z. Tang, G. Wei, Y. Zhu, Weighted sum rate maximization for OFDM-based cognitive radio systems. Telecommun. Syst. **42**, 77–84 (2009)
10. P. Wang, M. Zhao, L. Xiao, S. Zhou, J. Wang, Power allocation in OFDM-based cognitive radio systems, in *Proceedings of the IEEE Global Telecommunications Conference (GLOBECOM)*, November 2007, pp. 4061–4065
11. Q. Zhao, B.M.Sadler, A survey of dynamic spectrum access: signal processing, networking, and regulatory policy. IEEE Signal Process. Mag. **55**, 2294–2309 (2007)
12. A. Sultana, L. Zhao, X. Fernando, Power allocation using geometric water filling for OFDM-based cognitive radio networks, in *Proceedings of the IEEE International Vehicular Technology Conference (VTC-Fall)*, September 2016, pp. 1–5
13. D. Tse, P. Viswanath, *Fundamentals of Wireless Communications* (Cambridge University Press, Cambridge, 2004)
14. P. He, L. Zhao, S. Zhou, Z. Niu, Water-filling: a geometric approach and its application to solve generalized radio resource allocation problems. IEEE Trans. Wirel. Commun. **12**(7), 3637–3646 (2013)

Chapter 5
Resource Allocation for Cognitive Radio Systems with D2D Communication

5.1 Literature Review

There have been several burgeoning research efforts found in the literature [1–9] on the resource allocation schemes for CRNs. On the other hand, due to the benefit of D2D communication, International Mobile Telecommunications (IMT)—Advanced Standard systems, such as Long Term Evolution (LTE) and WiMax, allow D2D communication sharing the same radio resources with the cellular network to increase the spectral efficiency [10]. D2D communication is, in fact integrated into LTE-Advanced networks in [11, 12]. Radio resource allocation for D2D communication underlaying cellular networks are currently being extensively investigated by researchers in terms of spectral efficiency [13–18] and in terms of energy efficiency [19, 20].

From the literature survey, it is found that, most of the recent works focus on how to exploit D2D communication in cellular networks under different constraints. However, only a few efforts [10, 21–25] have been made to incorporate D2D communication with CR technology to jointly maximize the spectrum efficiency. In [10], the SUs' mode selection strategies are optimized in a cognitive cellular network with D2D communication. A mixed overlay-underlay spectrum sharing method is proposed in [21] for CR-assisted D2D communications in a cellular network. Two spectrum access policies (random and prioritized) are investigated in [22] for cognitive and energy harvesting-based D2D communication in cellular networks. A resource allocation scheme based on CR approach for D2D underlay multicast communication is proposed in [23] to improve system performance. An optimal power allocation algorithm is presented in [24] for cognitive D2D communication assisted by two-way relaying. The joint use of full-duplex relay and D2D communication are investigated in [25] for CRNs.

© The Author(s), under exclusive licence to Springer International Publishing AG, part of Springer Nature 2019
X. Fernando et al., *Cooperative Spectrum Sensing and Resource Allocation Strategies in Cognitive Radio Networks*, SpringerBriefs in Electrical and Computer Engineering, https://doi.org/10.1007/978-3-319-73957-1_5

5.2 System Model

A single cell downlink OFDM based cognitive cellular system with multiple cellular users as PUs and multiple D2D users as SUs is considered as shown in Fig. 5.1.

In spectral domain, we consider the side-by-side CR access model as shown in Fig. 5.2. It is assumed that the frequency bands with bandwidth B ($B_1, B_2, \ldots, B_K$ in Hz), are occupied by K PUs $(1, 2, \ldots, K)$. The unoccupied band that can be sensed by the M D2D users $(1, 2, \ldots, M)$ for possible transmission, is located on both sides of PU bands. It is possible for the D2D users to opportunistically utilize those unused spectrum by the of knowledge of the environment and cognition capability, to adapt to their radio parameters accordingly [26]. Let, $\mathcal{K}$ be the set for PUs where $\mathcal{K} = \{1, 2, \ldots, K\}$ and $\mathcal{M}$ be the set for D2D users where $\mathcal{M} = \{1, 2, \ldots, M\}$. The frequency band which is available for CR transmission, is divided into N subcarriers and each subcarrier occupies a bandwidth of ΔB Hz.

Each subcarrier transmits in the fading channel where channel gain is integrated by the effects of propagation path loss and shadowing. Thus, the D2D users need to detect the channel gain $h_{m,n}^{dd}$, $h_{m,k}^{dp}$ and $h_{k,m}^{pd}$, by the channel estimation mechanism

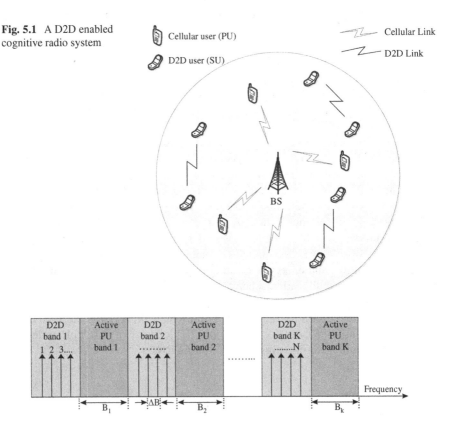

Fig. 5.1 A D2D enabled cognitive radio system

Fig. 5.2 Spectrum distribution of PUs and D2D users in OFDM based cognitive D2D system

before transmission. In practice, CSI on the D2D users own channel can be obtained via the classical channel estimation methods. However, the CSI from the D2D user to the PU is not directly available. This can be obtained in two steps. First the D2D user estimates the reversed channel from the PU to the D2D user. Then, under the assumption of channel reciprocity D2D to PU channel can be estimated by the D2D user. Here, $h_{m,n}^{dd}$ denotes the channel gain between the mth D2D user's transmitter and receiver; $h_{m,k}^{dp}$, is the channel gain between the mth D2D user's transmitter and the kth PU's receiver; and $h_{k,m}^{pd}$, is the channel gain between the kth PU's transmitter and the mth D2D user's receiver. The subscript of index n denotes the nth subcarrier. Table 5.1 provides a list of frequently used variables and abbreviations.

Table 5.1 List of frequently used variables and abbreviations

Group	Variable	Meaning
Index	k	Index of PUs, for $k = 1, 2, \cdots, K$
	m	Index of D2D users, for $m = 1, 2, \cdots, M$
	l	Index of other active D2D users, for $l = 1, 2, \cdots, M$
	n	Index of subcarriers, for $n = 1, 2, \cdots, N$
	B	Index of PUs bandwidth $B_1, B_2, \ldots, B_K$ in Hz
Set	$\mathcal{K}$	Set of PUs $\{1, 2, \cdots, K\}$
	$\mathcal{M}$	Set of D2D users $\{1, 2, \cdots, M\}$
	$\mathcal{N}$	Set of subcarriers $\{1, 2, \cdots, N\}$
	$\mathcal{N}_m$	Set of subcarriers allocated to the mth D2D user $\{1, 2, \cdots, N_m\}$
	$\bar{N}_m$	Number of subcarriers allocated to the mth D2D user
	λ	Set of subcarriers that exceeds the peak power constraint
	Ω	Set of subcarriers that need to be reallocated
Channel related term	$h_{m,n}^{dd}$	Channel gain between the mth D2D user's transmitter and receiver
	$h_{m,k}^{dp}$	Channel gain between the mth D2D user's transmitter and the kth PU's receiver
	$h_{k,m}^{pd}$	Channel gain between the kth PU's transmitter and the mth D2D user receiver
	$\Gamma_{m,n}$	Channel gain-to-noise ratio
	d_m	Link distance between the devices of D2D user
	η	Path attenuation factor
	σ_{AWGN}^2	Additive white Gaussian noise variance
Bandwidth related term	$d_{m,n}^{(k)}$	Spectrum distance between the nth subcarrier of the mth D2D user and the kth PU band
	ΔB	Bandwidth of each subcarrier in Hz
Power related term	$P_{m,n}$	Transmit power for the mth D2D user allocated to the nth subcarrier
	S_m	Total allocated power for the mth D2D user
	E_m	Extra power for the mth D2D user
	R_m	Remaining power for the mth D2D user that needs to be reallocated in the next round

(continued)

Table 5.1 (continued)

Interference related term	$\Upsilon_{m,n}^k$	Interference introduced by the nth subcarrier of the mth D2D user to the kth PU band
	$I_{m,n}^k$	Interference factor for the mth D2D user between the nth subcarrier and the kth PU band
	$\rho_{m,n}^k$	Interference introduced by the kth PU signal to the mth D2D user on the nth subcarrier
	$L_{l,n}$	Interference signal comes from other active D2D users
	$\mathscr{I}_{agg}$	Aggregated interference
Transmission rate related term	$\mathfrak{R}_{m,n}$	Transmission rate for the mth D2D user on the nth subcarrier
	$\bar{\mathfrak{R}}_{m,n}$	Transmission rate when peak power constraint is considered
Constraint Group	P_T	Total power constraint
	$P_{m,max}$	Maximum power constraint of the mth D2D user
	$\bar{P}_{m,n}$	Peak power constraint for the mth D2D user at the nth subcarrier
	I_{th}^k	Interference threshold of the kth PU band
	$\mathfrak{R}_{m,min}$	Minimum transmission rate requirements for the D2D users
	$x_{m,n}$	Binary decision variable of subcarrier allocation
Others	$\alpha, \beta, \gamma, \delta, \psi$	Lagrange multipliers
	$\Pi_{m,n}$	Proposed metric for subcarrier allocation
	w_1, w_2, w_3	Weighting factors for subcarrier allocation
Reference level	μ	Water level
	j^*	Water level step (highest step under the water)
Matrix	**X**	$M \times N$ matrix of subcarrier allocation
	P	$M \times N$ matrix of allocated power
	T	$M \times N$ matrix of allocated power when binary decision variable is considered
Abbreviations	PSD	Power spectrum density
	ASA	Adaptive subcarrier allocation
	PA	Power allocation
	GWF	Geometric water-filling
	GWFPP	Geometric water-filling with peak power constraint

Due to the coexistence of PU and D2D users in the side-by-side bands, the mutual interference is introduced. Now, the mutual interference introduced by the nth subcarrier of the mth D2D user to the kth PU band can be written as [27]

$$\Upsilon_{m,n}^{(k)} = |h_{m,k}^{dp}|^2 P_{m,n} T_s \int_{d_{m,n}^{(k)} - B_k/2}^{d_{m,n}^{(k)} + B_k/2} \left(\frac{sin\pi f T_s}{\pi f T_s} \right)^2 df \qquad (5.1)$$

where, $P_{m,n}$ is the transmit power for the mth D2D user allocated to the nth subcarrier, T_s is the symbol duration, $d_{m,n}^{(k)}$ represents the spectrum distance between the nth subcarrier of the mth D2D user and the kth PU band and B_k represents occupied bandwidth by the kth PU.

Let $I_{m,n}^{(k)} = |h_{m,k}^{dp}|^2 T_s \int_{d_{m,n}^{(k)} - B_k/2}^{d_{m,n}^{(k)} + B_k/2} \left(\frac{\sin \pi f T_s}{\pi f T_s} \right)^2 df$, be the interference factor for the mth D2D user between the nth subcarrier and the kth PU band. Then Eq. (5.1) can be rewritten as

$$\Upsilon_{m,n}^{(k)} = P_{m,n} \cdot I_{m,n}^{(k)}. \tag{5.2}$$

Now, with an ideal coding scheme, the transmission rate for the mth D2D user on the nth subcarrier, is given by the following formula

$$\Re_{m,n}(P_{m,n}, d_m) = \Delta B \log \left(1 + \frac{P_{m,n} |h_{m,n}^{dd}|^2 d_m^{-\eta}}{\mathscr{I}_{agg}} \right) \tag{5.3}$$

where, d_m represents the link distance between the devices of D2D user and η is the path attenuation factor. In (5.3), $\mathscr{I}_{agg} = \sigma_{AWGN}^2 + \sum_{k=1}^{K} \rho_{m,n}^k + \sum_{l=1, l \neq m}^{M} L_{l,n}$ denotes the aggregated interference, where σ_{AWGN}^2 denotes the additive white Gaussian noise (AWGN) variance, the second term be the interference introduced by the kth PU signal to the mth D2D user on the nth subcarrier and the third term represents the interference signal comes from other active D2D users.

Let $\Gamma_{m,n} = \frac{|h_{m,n}^{dd}|^2 d_m^{-\eta}}{\mathscr{I}_{agg}}$ represents the channel gain-to-noise ratio, then (5.3) can be rewritten as

$$\Re_{m,n}(P_{m,n}, d_m) = \Delta B \log \left(1 + P_{m,n} \Gamma_{m,n} \right). \tag{5.4}$$

In this paper, a transmission rate constraint associated with the maximum modulation order limitation on each subcarrier is considered. It leads to a peak power constraint ($\bar{P}_{m,n}$) for the mth D2D user on the nth subcarrier which can be written from (5.4) as

$$\bar{P}_{m,n} = \frac{2^{\bar{\Re}_{m,n}} - 1}{\Gamma_{m,n}} \tag{5.5}$$

where $\bar{\Re}_{m,n}$ is the corresponding transmission rate when $\bar{P}_{m,n}$ is considered.

Let $x_{m,n}$ be a binary decision variable of channel (subcarrier) allocation. If subcarrier n is allocated to D2D user m, then $x_{m,n}$ is equal to 1; otherwise it is equal to zero. Since each subcarrier is exclusively allocated to one D2D user, then

$$\sum_{m=1}^{M} x_{m,n} = 1, \quad n = 1, 2, \ldots, N. \tag{5.6}$$

Now, the total system transmission rate is given by

$$\Re(\mathbf{X}, \mathbf{P}) = \Delta B \sum_{m=1}^{M} \sum_{n=1}^{N} x_{m,n} \log\left(1 + P_{m,n}\Gamma_{m,n}\right) \tag{5.7}$$

where, $\mathbf{X}$ is a $M \times N$ matrix of subcarrier allocation indices $x_{m,n}$, and $\mathbf{P}$ is a $M \times N$ matrix of allocated power $P_{m,n}$.

5.3 Problem Formulation

Our objective of resource allocation (subcarrier and power allocation) is to maximize the total downlink transmission rate of the D2D system under both power and interference constraints along with the minimum rate requirements. Therefore, the optimization problem can be formulated as follows

$$\max_{\mathbf{X}, \mathbf{P}} \Delta B \sum_{m=1}^{M} \sum_{n=1}^{N} x_{m,n} \log\left(1 + P_{m,n}\Gamma_{m,n}\right) \tag{5.8}$$

subject to:

$$C1: \sum_{m=1}^{M} x_{m,n} = 1, \; x_{m,n} \in \{0, 1\}; \; \forall n \in \mathcal{N}$$

$$C2: \sum_{m=1}^{M} \sum_{n=1}^{N} x_{m,n} P_{m,n} \leq P_T, \; P_{m,n} \geq 0;$$

$$C3: \sum_{n=1}^{N} x_{m,n} P_{m,n} \leq P_{m,max}; \; \forall m \in \mathcal{M}$$

$$C4: x_{m,n} P_{m,n} \leq \bar{P}_{m,n}; \; \forall m \in \mathcal{M}, \; \forall n \in \mathcal{N}$$

$$C5: \sum_{k=1}^{K} \sum_{n=1}^{N} x_{m,n} \Upsilon_{m,n}^{(k)} \leq I_{th}; \; \forall m \in \mathcal{M}$$

$$C6: \sum_{n=1}^{N} x_{m,n} \Re_{m,n} \geq \Re_{m,min}; \; \forall m \in \mathcal{M}$$

where, constraint C1 depicts that each subcarrier can not be reused by more than one D2D user. Constraint C2, C3 and C4 denote total power constraint (e.g. P_T), maximum power constraint of each D2D user (e.g. $P_{m,max}$), and peak power constraint on each subcarrier (e.g. $\bar{P}_{m,n}$) respectively. Constraint C5 describes interference constraint where I_{th} denotes the total interference threshold by the K PU bands. Lastly, constraint C6 provides minimum transmission rate requirements for the D2D users.

The optimization problem in (5.8) is nonconvex due to binary decision variable $x_{m,n}$. This variable can be relaxed by applying a time sharing approach to allow any value in the interval $(0, 1]$ instead of $\{0, 1\}$ set. This approach allows multiple users to transmit on a certain subcarrier during a defined scheduling interval. Let $t_{m,n} = x_{m,n} P_{m,n}$, then the optimization problem in (5.8) can be rewritten as follows:

$$\max_{\mathbf{X}, \mathbf{T}} \Re(\mathbf{X}, \mathbf{T}) = \Delta B \sum_{m=1}^{M} \sum_{n=1}^{N} x_{m,n} \log \left(1 + \frac{t_{m,n}}{x_{m,n}} \Gamma_{m,n} \right) \tag{5.9}$$

subject to:

$$C1: \sum_{m=1}^{M} x_{m,n} = 1, \; x_{m,n} \in (0, 1]; \; \forall n \in \mathcal{N}$$

$$C2: \sum_{m=1}^{M} \sum_{n=1}^{N} t_{m,n} \leq P_T, \; t_{m,n} \geq 0;$$

$$C3: \sum_{n=1}^{N} t_{m,n} \leq P_{m,max}, \; \forall m \in \mathcal{M}$$

$$C4: t_{m,n} \leq \bar{P}_{m,n}; \; \forall m \in \mathcal{M}, \; \forall n \in \mathcal{N}$$

$$C5: \sum_{k=1}^{K} \sum_{n=1}^{N} t_{m,n} I_{m,n}^{(k)} \leq I_{th}; \; \forall m \in \mathcal{M}$$

$$C6: \sum_{n=1}^{N} x_{m,n} \Re_{m,n} \geq \Re_{m,min}; \; \forall m \in \mathcal{M}$$

Now, the problem in (5.9) is convex which is equivalent to the original problem in (5.8) when the condition on $x_{m,n}$ is relaxed for each D2D user m and subcarrier n. Finding the optimal pair $(x_{m,n}^*, t_{m,n}^*) = (x_{m,n}^*, x_{m,n}^* P_{m,n}^*)$ leads to the same solution as finding $(x_{m,n}^*, P_{m,n}^*)$.

The analytical solution of the problem in (5.9) is

$$P_{m,n}^* = \left[\frac{1 + \psi_m}{\ln \left(\alpha_n + \beta_m + \gamma_n + \sum_{k=1}^{K} \delta_k I_{m,n}^{(k)} \right)} - \frac{1}{\Gamma_{m,n}} \right]^+. \tag{5.10}$$

Proof See Appendix.

The $P_{m,n}^*$ denotes the optimal power allocation for the mth D2D user on the nth subcarrier when $m = m^*$. Thus, the indices of the subcarrier allocation matrix X will be

$$
x_{m,n} = \begin{cases} 1 & m = m^* \text{ for } n = 1, 2, \ldots, N \\ 0, & \text{otherwise.} \end{cases}
\tag{5.11}
$$

The power allocation strategy in (5.10) is indeed corresponding to the CWF strategy [28] and depends on the Lagrange multipliers associated with the per user power and interference constraints. By using the CWF, the optimal power allocation strategy can be calculated. However, several iterations may be involved in finding the optimal value and also the complexity of the optimal scheme is high. Thus, in this paper, GWF approach [29] is utilized to solve the CWF problem which is described in the following section. It has two advantages, (1) the geometric approach can compute the exact solution to the CWF, including the weighted case, with less computation without determining the water level through solving the non-linear system, (2) machinery of the proposed geometric approach can overcome the limitations of the CWF algorithm to include more stringent constraints.

5.4 Proposed Resource Allocation Scheme

Based on the analysis in the previous section, an efficient resource allocation scheme is proposed in this section to address the problem defined in (5.8). A two-stage approach is considered to solve (5.8) efficiently. Specifically, the resource allocation scheme is divided into two individual procedures: adaptive subcarrier allocation (ASA) and power allocation (PA). In the first stage, each subcarrier is assigned to one D2D user with the minimum value of the proposed metric that is adaptive in nature. In the second stage, power is allocated among all the D2D users to maximize the transmission rate. The procedure for the resource allocation scheme is illustrated in Fig. 5.3.

Fig. 5.3 The procedure for resource allocation scheme

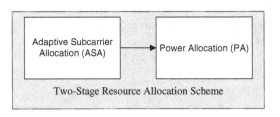

5.4.1 Adaptive Subcarrier Allocation (ASA)

Most of the recent works regarding subcarrier allocation strategy to the D2D users is based on greedy approach [15, 16]. In order to assign subcarriers to the D2D users, a proposed metric is utilized that is adaptive in nature. The proposed metric is related to the amount of power, interference and transmission rate.

Let $\mathcal{N}$ be the set of subcarriers where $\mathcal{N} = \{1, 2, \ldots, N\}$, $\mathcal{N}_m$ be the set of subcarriers allocated to the mth D2D user and $\bar{N}_m$ denotes the number of subcarriers allocated to the mth D2D user. Now, the transmission rate increment for one more subcarrier allocation to the mth D2D user can be written as

$$\Delta\mathfrak{R}_{m,n} = \mathfrak{R}_{m,n}\left(\frac{P_{m,n}}{\bar{N}_m + 1}, d_m\right) - \mathfrak{R}_{m,n}\left(\frac{P_{m,n}}{\bar{N}_m}, d_m\right), n \in \mathcal{N}_m. \qquad (5.12)$$

The power increment for transmitting one more increment in transmission rate $\Delta\mathfrak{R}$ on subcarrier n of the mth D2D user can be written as

$$\Delta P_{m,n} = \frac{(2^{\mathfrak{R}_{m,n}+\Delta\mathfrak{R}_{m,n}} - 2^{\mathfrak{R}_{m,n}})}{\Gamma_{m,n}}. \qquad (5.13)$$

The interference increment caused by subcarrier n of the mth D2D user for the kth PU can be written as

$$\Delta\Upsilon_{m,n}^{(k)} = \Delta P_{m,n} \cdot I_{m,n}^{(k)}. \qquad (5.14)$$

A metric composed of three parts for the nth subcarrier of the mth D2D user is described as [30]

$$\Pi_{m,n} = w_1\left(\frac{\Delta P_{m,n}}{P_{m,max} - P_{m,used}}\right) + w_2\left(\sum_{k=1}^{K}\frac{\Delta\Upsilon_{m,n}^k}{I_{th} - \Upsilon_{m,used}^k}\right) + w_3\left(\frac{\Delta\mathfrak{R}_{m,n}}{\mathfrak{R}_{m,used} - \mathfrak{R}_{m,min}}\right)$$

$$(5.15)$$

where, $P_{m,used}$ is the amount of power that has been utilized for the mth D2D user, $\Upsilon_{m,used}^k$ is the interference that has been initiated to the kth PU, and $\mathfrak{R}_{m,used}$ is the corresponding transmission rate of the mth D2D user. The first part consists of the amount of the power increment for transmitting one more increment in transmission rate and the amount of the unused power. The second part includes the amount of interference that is used for considering the interference constraints. Finally, the third part is associated with the minimum transmission rate requirements among the D2D users. Three non-negative parameters (w_1, w_2, w_3) are exploited as the weighting factors to make a trade-off between the power and the interference constraints together with the minimum rate requirements in the subcarrier allocation process. This is the reason why this subcarrier assignment strategy is named as adaptive subcarrier allocation (ASA).

Algorithm 1: Adaptive subcarrier allocation (ASA)

Input: The set $\mathscr{N} = \{1, 2, \dots, N\}$
1 initialization: $\mathscr{N}_m = \phi$, $\bar{N}_m = 0$, $\mathfrak{R}_{m,used} = 0$,
2 $P_{m,used} = 0$, $\Upsilon^k_{m,used} = 0$; $\forall m \in \mathscr{M}$, $\forall n \in \mathscr{N}$, $\forall k \in \mathscr{K}$
3 **for** $n \in \mathscr{N}$ **do**
4 $\quad$ compute $\Delta P_{m,n}$ using (5.13),
5 $\quad$ $\Delta\Upsilon^{(k)}_{m,n}$ using (5.14)
6 $\quad$ and $\Delta\mathfrak{R}_{m,n}$ using (5.12);
7 $\quad$ calculate $\Pi_{m,n}$ using (5.15);
8 $\quad$ $(m^*, n^*) = \arg\ \min\ \Pi_{m,n}$;
9 $\quad$ $\mathscr{N} = \mathscr{N} - \{n^*\}$, $\mathscr{N}_{m^*} = \mathscr{N}_m \cup \{n^*\}$, $\bar{N}_m = \bar{N}_m + 1$;
10 $\quad$ update $P_{m,used}$, $\Upsilon^k_{m,used}$, $\mathfrak{R}_{m,used}$
11 **end**

For a specific D2D user, a subcarrier n^* with the minimum value of the metric (5.15), is assigned to that selected D2D user m^*. A subcarrier with a lower value depicts that the power and interference increments for providing one more increment in transmission rate are relatively small compared to other subcarriers. The ASA strategy is well described in Algorithm 1.

5.4.2   Power Allocation (PA)

After subcarrier assignment described in the previous section, power allocation is performed for the selected subcarrier n^* of the D2D user m^*. For the selected subcarrier n^* of the D2D user m^*, the value of x_{m^*,n^*} becomes 1 and then the power allocation problem in (5.8) can be solved using GWF approach. Figure 5.4 gives an illustration of the GWF approach.

Instead of trying to determine the common water level μ (real nonnegative number) in CWF, the highest water level step, denoted by j^*, which is an integer number, is introduced to find the solutions for power allocation. Let $P(j)$ denotes the water volume above step j or zero, whichever is greater and the value of $P(j)$ can be found as:

$$P(j) = \left[P_T - \left\{ \sum_{n=1}^{j-1} \left(\frac{1}{\Gamma_{m,j}} - \frac{1}{\Gamma_{m,n}} \right) \right\} \right]^+ ; \forall j \in \mathscr{N}_m \qquad (5.16)$$

where $\frac{1}{\Gamma_{m,n}}$ is the "step depth" of the nth stair. Due to the definition of $P(j)$ being the power (water volume) above step j, it cannot be a negative number. Therefore

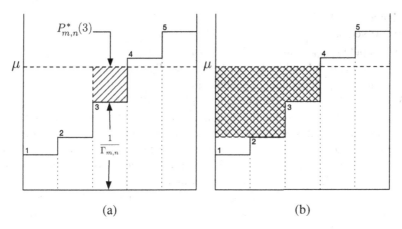

Fig. 5.4 Illustration for the GWF algorithm (**a**) Illustration of water level step $j^* = 3$, allocated power for the third step $P^*_{m,n}(3)$, and step depth is $\frac{1}{\Gamma_{m,n}}$ and (**b**) Illustration of $P(j)$ (shadowed area, representing the total power above step j) when $j = 2$

we use $\{.\}^+$ in (5.16) to assign 0 to $P(j)$ if the result inside the bracket is negative. The corresponding geometric meaning is that the $j^{*^{th}}$ level is above water.

According to [29], the explicit solution for power allocation is:

$$P_{m,n} = \begin{cases} P_{m,j^*} + \left(\frac{1}{\Gamma_{m,j^*}} - \frac{1}{\Gamma_{m,n}}\right) & 1 \le n \le j^* \\ 0, & j^* < n \le N_m; \end{cases} \quad (5.17)$$

where the water level step j^* is given as

$$j^* = \max\{j \,|\, P(j) > 0, 1 \le j \le N_m\} \quad (5.18)$$

and the power level for this step is

$$P_{m,j^*} = \frac{1}{j^*} P(j^*). \quad (5.19)$$

In order to allocate power to the selected subcarriers for each D2D user, first we perform geometric water-filling with peak power constraint (GWFPP) [29] for all the subcarriers and calculate $\{P_{m,n}\}$ using (5.17)–(5.19). Now, let λ is defined by the set $\{n \,|\, P_{m,n} > \bar{P}_{m,n}, n \in \mathcal{N}_m\}$. If λ is an empty set, then $P_{m,n}$ will be the output, otherwise, $P_{m,n}$ will be set by the individual peak power $\bar{P}_{m,n}$. Everytime we need to update the set $\mathcal{N}_m$ and the total power P_T at the end of each iteration. Since the finite set $\mathcal{N}_m$ is getting smaller and smaller until the set λ is empty. The required steps to apply the algorithm GWFPP is summarized below:

Algorithm GWFPP

Input: vector $\{\frac{1}{\Gamma_{m,n}}\}, \{\bar{P}_{m,n}\} \; \forall n \in \mathcal{N}_m$ and P_T.

1) Compute $\{P_{m,n}\}$ using (5.17), (5.18) and (5.19).

2) The set λ is defined by the set $\{n | P_{m,n} > \bar{P}_{m,n}, n \in \mathcal{N}_m\}$. If λ is the empty set, output $\{P_{m,n}\}_{n=1}^{N_m}$; else, $P_{m,n} = \bar{P}_{m,n}$, as $n \in \lambda$.

3) Update $\mathcal{N}_m$ and P_T. Then return to 1) of the GWFPP.

The next step is to compute the summation of allocated power for all D2D users. Let S_m be the total allocated power for the mth D2D user, E_m be the extra power for the mth D2D user when S_m exceeds the device's maximum power ($P_{m,max}$) and R_m be the remaining power for the mth D2D user that needs to be reallocated in the next round. In order to satisfy device's maximum power constraint, three cases are considered: (1) $S_m = P_{m,max}$, (2) $S_m > P_{m,max}$ and (3) $S_m < P_{m,max}$. For the first case, after checking the interference and minimum rate requirement constraints, we can have the allocated power vector for those D2D users directly. For the second case, we first perform GWFPP with the corresponding $P_{m,max}$ and then, check the interference and minimum rate requirement constraints to get their power allocation vectors. In this case, we also calculate all the E_m that needs to be reallocated in the next round. Finally, for the third case, in order to perform GWFPP less number of times, we start with those D2D users where we can allocate the remaining power most while satisfying the required constraints. In each case, if any violation of the interference or minimum rate constraint happens, this algorithm creates a set of Ω for those subcarriers and the amount of the R_m that needs to be reallocated in the next round. The process continues until all the reallocation being completed while satisfying all the constraints. The required steps for performing the power allocation are described in Fig. 5.5 and the detailed algorithm is well depicted in the Algorithm 2 description.

5.5 Performance Evaluation

In order to evaluate the performance of the subcarrier assignment and power allocation algorithm for D2D communications, we simulate a multiuser single cell system with radius 100 m where BS is located in the center of the cell. All cellular users (perform as PUs) and D2D users (perform as SUs) are uniformly distributed randomly within the cell. The distance between D2D pairs varies depending on their relative position in the cell. All other simulation parameters are considered according to Table 5.2. The simulation results are evaluated over different realization of PUs and D2D users locations, interference conditions and channel gains. Average transmitted data rates for different algorithms under consideration are obtained from several independent simulation runs.

In order to evaluate the performance of the proposed algorithm, we compared our algorithm with the classical schemes: uniform power loading, water-filling schemes [28], and ladder/triangular profile power allocation [31, 32]. In the uniform

---

Algorithm 2: Power allocation (PA)

Input: vector $\{P_{m,max}\} \forall m \in \mathscr{M}$, P_T;
1 initialization;
2 $A = \{m | m \in \mathscr{M}\}$, $B = \{n | n \in \mathscr{N}_m\}$
3 $C = \{m | \text{sorted D2D user}\}$, $\Omega = \phi$;
4 **for** $n \in B$ **do**
5 | compute $\{P_{m,n}\}$ using **GWFPP** with P_T
6 **end**
7 **for** $m \in A$ **do**
8 | calculate total allocated power, $S_m = \sum_{n=1}^{N_m} P_{m,n}$
9 | **if** $S_m = P_{m,max}$ **then**
10 | | **Interference-Rate Check**$(\Upsilon_m, \Re_m)$;
11 | **end**
12 | **if** $S_m > P_{m,max}$ **then**
13 | | Extra power, $E_m = S_m - P_{m,max}$;
14 | | compute $\{P_{m,n}\}$ using **GWFPP** with $P_{m,max}$;
15 | | **Interference-Rate Check**$(\Upsilon_m, \Re_m)$;
16 | **end**
17 | **if** $S_m < P_{m,max}$ **then**
18 | | calculate $\{P_{m,max} - S_m\}$;
19 | **end**
20 **end**
21 sort $\downarrow \{P_{m,max} - S_m\}$;
22 **for** $m \in C$ **do**
23 | total power need to be reallocated, $R_m = S_m + E_m$;
24 | $\mathscr{N}_m = \mathscr{N}_m \cup \{n\}, n \in \Omega$;
25 | **if** $R_m \leq P_{m,max}$ **then**
26 | | compute $\{P_{m,n}\}$ using **GWFPP** with R_m;
27 | | **Interference-Rate Check**$(\Upsilon_m, \Re_m)$;
28 | **else**
29 | | Extra power, $E_m = S_m - P_{m,max}$;
30 | | compute $\{P_{m,n}\}$ using **GWFPP** with $P_{m,max}$;
31 | | **Interference-Rate Check**$(\Upsilon_m, \Re_m)$;
32 | **end**
33 **end**
34 **return** $\{P_{m,n}\}$
35 **Function** `Interference-Rate Check`$(\Upsilon_m, \Re_m)$:
36 | calculate total interference, $\Upsilon_m = \sum_{n=1}^{N_m} P_{m,n} I_{m,n}^{(k)}$;
37 | calculate total transmission rate, $\Re_m = \sum_{n=1}^{N_m} \Re_{m,n}$
38 | **if** $(\Upsilon_m \leq I_{th})$ && $(\Re_m \geq \Re_{min})$ **then**
39 | | output: $\{P_{m,n}\}$;
40 | | $E_m = 0$;
41 | **end**
42 | If($!\mathscr{N}_m$)
43 | $\Omega \longleftarrow \mathscr{N}_m \setminus n, E_m = \sum_{n=1}^{N_m} P_{m,n}$
44 | **return** $\{P_{m,n}\}, E_m, \Omega$

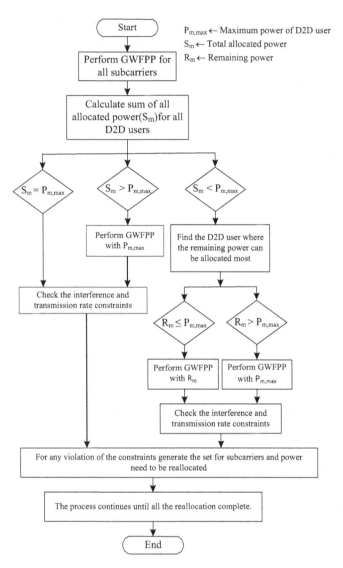

Fig. 5.5 Flow chart of power allocation algorithm

power loading scheme, uniform power is loaded into each subcarrier. With the total power as the power constraint, the power profile follows the water-filling scheme as in [28]. In ladder profile scheme [31], power is distributed in such a fashion so that the subcarriers that are adjacent to the PU bands, are given less power where as the subcarriers that are far away to the PU bands, are given more power. According to the triangular power allocation scheme in [32], power allocated to the subcarriers near the PUs is small and gradually increase as move towards the middle. Thus

Table 5.2 Simulation parameters

Parameters	Values
Total no. of D2D users	5–35
Total no. of PUs	12–24
Total no. of subcarriers	5–50
Subcarrier bandwidth (ΔB)	1.5 kHz
Total bandwidth (B)	10 kHz
The value of T_s	1 µs
Path loss factor	4
The value of δ_d^2	10^{-12}
Max D2D T_x power	0.1 W
Max PU T_x power	0.25 W

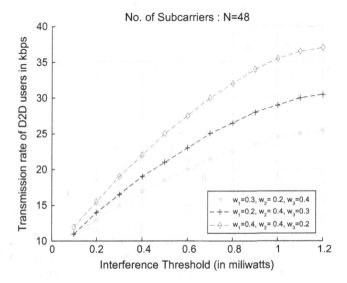

Fig. 5.6 Illustration of the behavior for the proposed subcarrier assignment with different values of the weighting factors (w_1, w_2, w_3)

maximum power is allocated to the middle most subcarrier. Since the idea behind the schemes (ladder profile and triangular power allocation) is the same, thus we name these schemes as ladder/triangle scheme.

The behavior of the proposed subcarrier assignment strategy with different values of the weighting factors in (5.15) is illustrated in Fig. 5.6. The effect of the different weighting factors (w_1, w_2, w_3) is revealed in the performance curves. The simulation results show that an appropriate value of the weighting factors for the three parts in the proposed indicator can achieve superior performance. We select the values of the weighting factors (w_1, w_2, w_3) to be $(0.4, 0.4, 0.2)$ in the following simulation, since this combination returns the highest sum rate of the D2D uses in the simulation range.

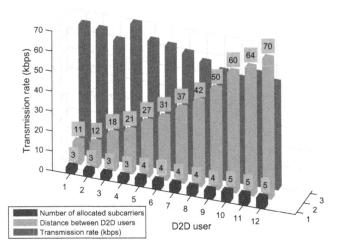

Fig. 5.7 Effect of the distance between the D2D users on subcarrier allocation

Figure 5.7 shows the effect of the distance between the D2D users on subcarrier allocation where the minimum rate requirement is 50 kbps. Here, the first row represents the number of allocated subcarriers, the second row denotes the distance between the D2D users and the third row shows the transmission rate. It can be observed from Fig. 5.7 that the distance between the D2D user plays an important role on the total number of subcarriers allocated to one D2D user. To satisfy the minimum rate requirements of the D2D users, closely located D2D users require less number of subcarriers compared to the distantly located D2D users. Figure 5.7 depicts the scenario where D2D users located within 20 m need three subcarriers to satisfy the minimum rate requirement whereas D2D users located within 70–80 m range need five subcarriers to satisfy the minimum rate requirement.

In Fig. 5.8, we plotted the transmission rate of D2D users versus the interference threshold introduced to the PUs band for different schemes. As we see, the transmission rate for all the schemes increases as the interference threshold increases. Also the transmission rate achieved by the proposed scheme outperforms other three existing schemes. The reason is that the proposed scheme maintains both the power and interference constraints at all stages of the operation. The transmission rate of the uniform power loading scheme is significantly lower than the other schemes due to the violation of both of these constraints.

Figure 5.9 shows the transmission rate achieved by the D2D users versus the number of subcarriers for the different schemes. It is obvious that as the number of subcarriers increases, the proposed scheme provides better transmission rate for D2D users than the other existing methods. The reason is that our proposed ASA scheme is adaptive in nature due to the three different parts in the used metric.

Figure 5.10 presents the interference introduced to the PU band versus total power budget for different schemes. As the total power budget increases, the

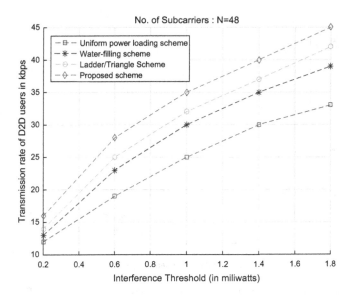

Fig. 5.8 Transmission rate of D2D users versus interference threshold to the PUs for different schemes

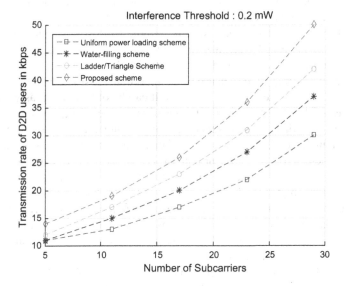

Fig. 5.9 Transmission rate of D2D users versus number of subcarriers with fixed interference threshold for different schemes

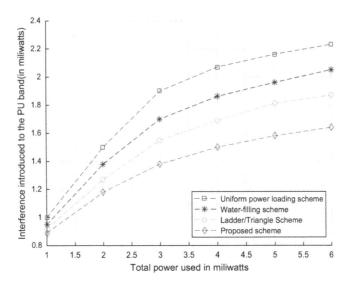

Fig. 5.10 Total power budget versus interference introduced to the PU band for different schemes

interference generated to the PU band by the uniform power loading and water-filling schemes are severe due to not taking the interference constraints into account. On the other hand, our proposed scheme is able to load power into multi-D2D users providing increased transmission rate while always keeping the interference introduced to the PU band below a specified threshold.

Figure 5.11 illustrates the scenario of the total transmission rate versus the number of D2D users for the different schemes. It can be observed that as we increase the number of D2D users, the total transmission rate of our proposed scheme outperforms the other existing schemes while maintaining different constraints. When the number of D2D users remain small, the total transmission rate graph for all the schemes follow the linear pattern with a larger slope. However, for large number of D2D users, the total transmission rate graph for all the schemes become little flat with a smaller slope. This is obvious that as we increase the number of D2D users, the interference constraint become dominant in this region and the total transmission rate does not increase as the number of D2D users increase. However, in both region (linear or little flat), our proposed scheme provides superior performance than the other schemes.

Figure 5.12 investigates the performance of different algorithms for different minimum rate requirements. Here, three different groups (e.g. G1, G2 and G3) are categorized for three different minimum rate requirement values (e.g. 10, 50 and 100 kbps) to evaluate the algorithms performance in terms of spectral efficiency. Spectral efficiency is defined as the total transmission rate divided by the bandwidth of the spectrum dedicated to the D2D user with unit bits/s/Hz. The comparison result presents that due to an increase in the minimum rate requirement, for the

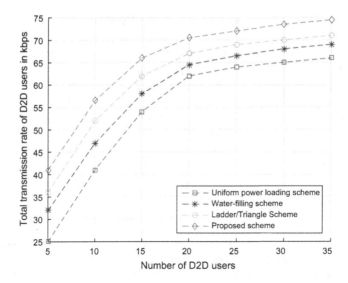

Fig. 5.11 Total transmission rate versus number of D2D users for the different schemes

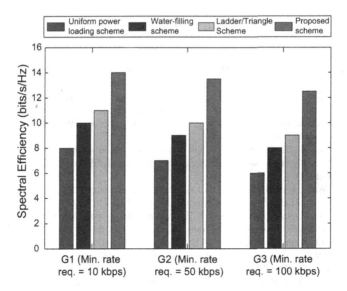

Fig. 5.12 Performance of different schemes for different minimum rate requirements

Table 5.3 Improvement (in %) of the proposed scheme on spectral efficiency compared to other schemes

Groups	Other schemes	Improvement (%)
G1 (Min. rate req. 10 kbps)	Uniform power loading	43
	Water-filling	29
	Ladder/triangle	22
G2 (Min. rate req. 50 kbps)	Uniform power loading	48
	Water-filling	33
	Ladder/triangle	26
G3 (Min. rate req. 100 kbps)	Uniform power loading	52
	Water-filling	36
	Ladder/triangle	28

same constraints, the spectral efficiency decreases. This is because less number of D2D users are admitted due to high rate requirement threshold and stringent power and interference constraints. However, still the proposed scheme outperforms the other schemes which is well depicted in Table 5.3.

References

1. G.I. Tsiropoulos, O.A. Dobre, M.H. Ahmed, K.E. Baddour, Radio resource allocation techniques for efficient spectrum access in cognitive radio networks, in *IEEE Communications Surveys and Tutorials*, First quarter, vol. 18, no. 1 (2016), pp. 824–847
2. X. Li, N. Zhao, Y. Sun, F.R. Yu, Interference alignment based on antenna selection with imperfect channel state information in cognitive radio networks. IEEE Trans. Veh. Technol. **64**(7), 5497–5511 (2016)
3. N. Zhao, F.R. Yu, H. Sun, M. Li, Adaptive power allocation schemes for spectrum sharing in interference-alignment-based cognitive radio networks. IEEE Trans. Veh. Technol. **65**(5), 3700–3714 (2016)
4. H. Xie, B. Wang, F. Gao, S. Jin, A full-space spectrum-sharing strategy for massive MIMO cognitive radio. IEEE J. Sel. Areas Commun. **34**(10), 2537–2549 (2016)
5. E. Bedeer, O.A. Dobre, M.H. Ahmed, K.E. Baddour, Rate – interference tradeoff in OFDM-based cognitive radio systems. IEEE Trans. Wirel. Commun. **64**(9), 4292–4298 (2015)
6. E. Bedeer, O. Amin, O.A. Dobre, M.H. Ahmed, K.E. Baddour, Energy-efficient power loading for OFDM-based cognitive radio systems with channel uncertainties. IEEE Trans. Veh. Technol. **64**(6), 2672–2677 (2015)
7. M. El-Absi, M. Shaat, F. Bader, T. Kaiser, Interference alignment with frequency-clustering for efficient resource allocation in cognitive radio networks. IEEE Trans. Wirel. Commun. **14**(12), 7070–7082 (2015)
8. E. Bedeer, O.A. Dobre, M.H. Ahmed, K.E. Baddour, A multiobjective optimization approach for optimal link adaptation of OFDM-based cognitive radio systems with imperfect spectrum sensing. IEEE Trans. Wirel. Commun. **13**(4), 2339–2351 (2014)
9. E. Bedeer, O.A. Dobre, M.H. Ahmed, K.E. Baddour, Joint optimization of bit and power loading for multicarrier systems. IEEE Wirel. Commun. Lett. **2**(4), 447–450 (2013)

10. P. Cheng, L. Deng, H. Yu, Y. Xu, H. Wang, Resource allocation for cognitive networks with D2D communication: an evolutionary approach, in *Proceedings of IEEE Wireless Communications and Networking Conference* (2012), pp. 2671–2676

11. J. Liu, S. Zhang, N. Kato, H. Ujikawa, K. Suzuki, Device-to-device communications for enhancing quality of experience in software defined multi-tier LTE-A networks. IEEE Netw. **29**(4), 46–52 (2015)

12. J. Liu, N. Kato, J. Ma, N. Kadowaki, Device-to-device communication in LTE-Advanced networks: a survey. IEEE Commun. Surv. Tutorials **17**(4), 1923–1940 (2015)

13. R. Yin, C. Zhong, G. Yu, Z. Zhang, K.K. Wong, X. Chen, Joint spectrum and power allocation for D2D communications underlaying cellular networks. IEEE Trans. Veh. Technol. **65**(4), 2182–2195 (2016)

14. C. Xu, L. Y. Song, Z. Han, Q. Zhao, X. Wang, B. Jiao, Efficiency resource allocation for device-to-device underlay communication systems: a reverse iterative combinatorial auction based approach. IEEE J. Sel. Areas Commun. **31**(9), 348–358 (2013)

15. H.H. Esmat, M.M. Elmesalawy, I.I. Ibrahim, Adaptive resource sharing algorithm for device-to-device communications underlaying cellular networks. IEEE Commun. Lett. **20**(3), 530–533 (2016)

16. W. Zhao, S. Wang, Resource sharing scheme for device-to-device communication underlaying cellular networks. IEEE Trans. Commun. **63**(12), 4838–4848 (2015)

17. D. Zhu, J. Wang, A.L. Swindlehurst, C. Zhao, Downlink resource reuse for device-to-device communications underlaying cellular networks. IEEE Signal Processing Lett. **21**(5), 531–534 (2014)

18. D. Feng, L. Lu, Y. Yuan-Wu, G.Y. Li, G. Feng, S. Li, Device-to-device communications underlaying cellular networks. IEEE Trans. Commun. **61**(8), 3541–3551 (2013)

19. Y. Jiang, Q. Liu, F. Zheng, X. Gao, X. You, Energy-efficient joint resource allocation and power control for D2D communications. IEEE Trans. Veh. Technol. **65**(8), 6119–6127 (2016)

20. X. Chen, R.Q. Hu, J. Jeon, G. Wu, Energy efficient resource allocation for D2D communication underlaying cellular networks, in *Proceedings of IEEE International Conference on Communications* (2015), pp. 2943–2948

21. M.G. Khoshkholgh, Y. Zhang, K.-C. Chen, K.G. Shin, S. Gjessing, Connectivity of cognitive device-to-device communications underlying cellular networks. IEEE J. Sel. Areas Commun. **33**(1), 81–99 (2015)

22. A.H. Sakr, E. Hossain, Cognitive and energy harvesting-based D2D communication in cellular networks: stochastic geometry modeling and analysis. IEEE Trans. Commun. **63**(5), 1867–1880 (2015)

23. X. Wu, Y. Chen, X. Yuan, M.E. Mkiramweni, Joint resource allocation and power control for cellular and device-to-device multicast based on cognitive radio. IET Commun. **8**(16), 2805–2813 (2014)

24. L. Wang, H. Jin, X. Ji, Y. Li, M. Peng, Power allocation optimization for D2D communication underlaying cognitive full duplex relay networks, in *Proceeding of IEEE Conference on Wireless Communications, Networking and Mobile Computing (WiCOM)* (2016), pp. 1–6

25. L. Wang, H. Jin, X. Ji, Y. Li, M. Peng, Power allocation for cognitive D2D communication assisted by two-way relaying, in *IEEE Symposium on Microwave, Antenna, Propagation and EMC Technologies for Wireless Communications (MAPE)* (2013), pp. 1–6

26. A.H. Sakr, H. Tabassum, E. Hossain, D.I. Kim, Cognitive spectrum access in device-to-device-enabled cellular networks. IEEE Commun. Mag. **53**(7), 126–133 (2015)

27. T. Weiss, J. Hillenbrand, A. Krohn, F.K. Jondral, Mutual interference in OFDM-based spectrum pooling systems, in *Proceedings of IEEE Vehicular Technology Conference (VTC 2004-Spring)*, vol. 4 (2004), pp. 1873–1877

28. D.P. Palomar, J.R. Fonollosa, Practical algorithms for a family of water-filling solutions. IEEE Trans. Signal Process. **53**(2), 686–695 (2005)

29. P. He, L. Zhao, S. Zhou, Z. Niu, Water-filling: a geometric approach and its application to solve generalized radio resource allocation problems, in *IEEE Transactions on Wireless Communications* (2013), pp. 3637–3646

30. A. Sultana, L. Zhao, X. Fernando, Efficient resource allocation in device-to-device communication using cognitive radio technology. IEEE Trans. Veh. Technol. **PP**(99), 1–12 (2017)
31. G. Bansal, M.J. Hossain, V.K. Bhargava, Optimal and suboptimal power allocation schemes for OFDM-based cognitive radio systems. IEEE Trans. Wirel. Commun. **7**(11), 4710–4718 (2008)
32. S. Chakraborty, P. Dhanuka, A. Kumar, S.P. Maity, Subcarrier and power allocation schemes for multiuser OFDM-based cognitive radio systems, in *Proceeding of IEEE National Conference on Communications (NCC)* (2013), pp. 1–5

Chapter 6
Conclusion and Future Work

6.1 Conclusion

Spectrum crunch escalates as wireless communication solutions, both human and machine centric, being deployed overwhelmingly and require more and more bandwidth. In the realm of such severe spectrum scarcity, a sustainable solution for the spectrum crunch is essential. The CRN solution, that enables intelligent spectrum sharing and dynamic spectrum access could provide a serious long term solution. This is further facilitated by the dominance of software control in wireless systems, both at the transceiver level and the network level.

This book discusses certain key issues upon which a practical CRN may be built. The book addresses few main challenges such as reliable spectrum sensing, an essential part of interweave distributed CRN deployment and optimal resource management considering D2D communications as a case study.

In detail, Chap. 1 presents an overview of cognitive radio networks and defines overlay, underlay and interweave type CRNs. This chapter also discusses the advantages of implementing cooperative spectrum sensing to achieve sensing diversity gain. An overview of the centralized and distributed network structures are given. Various spectrum sensing and diversity combining techniques are also discussed.

Chapter 2 analyzes relay based cooperative spectrum sensing performance over fading channels through intensive analytical and simulated evaluations. The obtained results show the importance of including relaying link impairments and the right combining techniques in the evaluation of CSS performance. It has been shown that multipath fading is strongly affects the reliability of primary user detection, causing large deviations from those results obtained using AWGN-based models. The derived expressions for the average false alarm probability and the average detection probability, introduced in this chapter, could lead to intuitive CRN design

© The Author(s), under exclusive licence to Springer International Publishing AG, part of Springer Nature 2019
X. Fernando et al., *Cooperative Spectrum Sensing and Resource Allocation Strategies in Cognitive Radio Networks*, SpringerBriefs in Electrical and Computer Engineering, https://doi.org/10.1007/978-3-319-73957-1_6

guidelines. They can be used to determine the energy threshold value, the minimum number of collected energy samples, and the maximum transmission power that meet a given detection accuracy over non-identical fading channels. Generally, our analysis reveals that: (1) cooperation allows independently faded radios to collectively achieve robustness to serve fades, (2) multipath fading on relaying channels yields similar performance degradations as multipath fading on sensing channels, (3) a small number of radios are enough to achieve practical detection levels, and (4) practical performance measures for dealing with fading strongly depend on the target probability of detection.

Chapter 3 provides a taxonomy for the centralized and distributed resource allocation process in CRNs for both infrastructure based or ad-hoc type CRNs. The resource allocation is a multivariate optimization problem that is often solved considering the throughput as the objective function and factors like transmission power, maximum transmission delay, allowable interference, QoS and fairness as the constrains. This optimization problem can be solved using classical optimization algorithms or using heuristic approaches.

In this book more OFDM, that has been widely accepted as the de-facto standard for 5G networks is used as the foundation.[1] Although different optimization strategies are possible, we focus on a modified GWF algorithm applied in general OFDM based cognitive radio systems considering per subchannel power constraint and individual peak power constraint aiming to maximize the throughput.

For this purpose, an efficient power and subcarrier allocation algorithm, named as IGPP is discussed. This algorithm is optimized in such a way to consider both individual subcarrier peak power constraint and per subchannel (group of subcarriers) power constraint. Simulation results are provided to illustrate the effectiveness of the algorithm. The proposed algorithm allows better utilization of the available power resources, thus maximizes the throughput than some other existing algorithms. The algorithm allocates more power to wireless channels that have less fading, hence saving power resources. Also a weight factor is considered for each channel to further fine tune the optimization.

Then as a case study, it is studied how efficiently D2D communications can be employed with a CR approach. Here, again an OFDM based cognitive cellular network with D2D communications has been modeled and analyzed using Lagrange formulation. The adaptive subcarrier allocation is optimized in terms of power, interference and transmission rate. The modified GWF algorithm is tested to maximize the total downlink transmission rate of the D2D system under both power and interference constraints along with minimum rate requirements. This scheme outperforms the existing schemes as revealed in the simulation results.

In a nutshell, the objective of this short book is to bring awareness on the key issues in designing a reliable high performance interweave CRN with cooperative

---

[1]Although Non Orthogonal Medium Access (NOMA) type algorithms are discussed for 5G networks, their performances are not yet proven and they require a significantly differential power transmission which may not be practical.

spectrum sensing. It has been shown that both realistic sensing and reporting channels have to considered for reliable cooperative spectrum sensing. Then, the need for optimal resource allocation is discussed and few algorithms proposed by the authors are used to perform an optimal resource allocation as a case study.

6.2 Future Research Directions

In this section, the challenges in cooperative spectrum sensing and RA methods are discussed and also the future research directions and opportunities are outlined.

6.2.1 Spectrum Sensing Issues

The existing RA schemes are developed under the assumption of having perfect spectrum sensing. However, this assumption may not be true and there may be errors in spectrum sensing which may direct to inefficient resource allocation and interference with PUs. Therefore, discovery process of spectrum need to be carried out more carefully because it is influenced by three main issues:

- The hidden transmitter
- The exposed transmitter and
- The hidden receiver.

The hidden transmitter issue has been resolved by carrying out sensing operation at both transmitter and receiver ends, but there are still no satisfactory solutions for the latter issues. A CR user should be capable of discovering the existence of a neighboring primary receiver to resolve these issues. Hence, feasible solutions for this purpose are yet to be investigated. Furthermore, spectrum sensing performance is limited by hardware and physical constraints. For instance, SUs with a single transceiver cannot transmit and sense simultaneously. Moreover, users usually only observe a partial state of the network to limit sensing overhead. There is a fundamental trade-off between the undesired overhead and spectrum holes detection effectiveness: the more bands are sensed, the higher the number and quality of the available resource. Thus this problem needs to be further investigated to improve spectrum sensing effectiveness. Lastly, sensing errors due to false detection and miss detection in PU signal sensing are ignored in many cases in the literature. Accurate models that take both false alarm and mis-detection probabilities into account, need to be devised. Therefore, the simplified ON/OFF model for PU traffic may not be a proper choice for practical environment where PUs may be cellular or wireless sensors. To improve the probability of detection with less sensing errors, cooperative techniques have been proposed [1]. However, many aspects of cooperative sensing need to be still investigated for better solutions [2]. Future RA schemes have to incorporate more complex considerations such as the impact of correlation among sensing channels, imperfectness of reporting channels etc. to develop more efficient cooperative sensing schemes [3]. Furthermore, trade-off between sensing action and throughput as well as other applications such as multimedia application over CRNs needs to be further investigated.

6.2.2 Channel Models

The majority of the proposed RA algorithms have adopted a Rayleigh fading channel model. Few works considered Nakagami-m channel model [4]. However, it is suggested to test the existing algorithms and propose new algorithms for different channel models. Particularly, the environment where the CR entities communicate whether urban, sub-urban, rural, indoor, or outdoor will affect the wireless channel characteristics. For example, the deployment of a secondary network in a rural area with low building density increases the line-of-sight probability, and thus Rician fading channels would be more appropriate. Altering the adopted channel model will definitely affect the analytical performance results and also could have an effect on the algorithms optimality. Furthermore, more efforts considering non-i.i.d channels is encouraged [5].

6.2.3 Cross-Layer Design Approaches and Security Issues

In CRNs, superior interaction is needed between different layers of protocol stacks in order to attain desired goals and performance in terms of radio resource management, QoS provisioning, security and other network objectives. The idea behind the Cross-layer design [6] approach refers to devise the protocol stack by exploiting the information exchange and reliance between different protocol layers to acquire superior performance. But at the same time, it will also be susceptible to cross-layer attacks which may happen due to malevolent operations executed at one layer that could cause security violations at another layer [7]. CRNs inherently require greater interaction between different layers of the protocol stack. Therefore, cross-layer attacks and security related issues [8] including jamming attack and mitigation, selfish behavior in collaborative sensing and misbehavior in detection, physical layer and MAC layer in security, and the modeling and detection of insider attacks should be given special attention. A cross-layer security mechanism that incorporates recent advances on security threats/attacks and countermeasures, need to be investigated in RA design in order to permit a reliable and secure environment for CRNs [9]. Finally, the application of artificial intelligence (AI) techniques can be included to tackle security challenges for dynamic spectrum access [10].

6.2.4 Spectrum Mobility Issues

Spectrum mobility is also an essential issue in RA design aspect. Studies in the literature incorporate several design issues such as PU detection, handoff decision, target channel selection and spectrum handoff strategy [11]. Spectrum sensing speed and precision that greatly influence PU detection event, can be increased by cooperative sensing [12]. Handoff decision is another important issue which can cause harmful interference to PU. This can be greatly improved by using proper handoff algorithm e.g. fuzzy logic based algorithm [13]. Appropriate target channel selection approaches in the literature include having a backup channel, target channel availability prediction under partial sensing scheme and selection utilizing historical data. PU traffic modeling taking PU mobility [14] into account needs careful attention in order to intelligently select the target channel by SUs. Several prediction and estimation schemes [15], like hidden Markov models [16], neural networks [17], Bayesian inference [18] and, autoregressive model [19] can be adopted for better realization of PU activity. Proper spectrum handoff strategies need to be chosen that are adaptive in nature according to PU traffic. In order to minimize delay in handoff event, simultaneous data transmission in multi-channel CRNs need to be performed. Effective contention resolution for multiple SUs during handoff need to be carried out for successful handoff. Cross-layer approach for link maintenance is necessary to effectively address mobility issues in physical, MAC and network layers. Finally an integrated handoff management process [20] is necessary to improve network performance.

6.2.5 Multi-Hop Scenario

Dynamic spectrum access in multi-hop domain incorporates several challenges in RA schemes. Several studies [21, 22] in the literature tried to bring up research issues regarding RA design for multi-hop CRNs which can be summarized as:

1. Control channel establishment and management scheme without a predefined dedicated control channel,
2. Transceiver synchronization,
3. CTS timeout and problems in decoding CTS,
4. Multi-channel hidden terminal problem,
5. Hidden incumbent node problem,
6. Number of transceivers,
7. Coordination of spectrum sensing and accessing decision making,
8. Radio frequency heterogeneity,
9. Group communication, and
10. MAC layer authentication.

Some other attributes like the effect of channel state information in spectrum access, QoS guarantee, concern in fairness aspect, PU protection also need careful attention and further investigation in multi-hop environment.

6.2.6 QoS Management in CR Environment

Spectrum utilization and network capacity of CRNs can be increased by dynamic spectrum supervision. However, this imposes several challenges in QoS management [23]. Users can exploit available spectrum successfully captured by a CR but accurate information needs to be forwarded to the application to amend traffic features and user requirements. Once the available frequency bands are characterized, the most suitable spectrum bands can be selected by taking into account the spectrum characteristics and QoS criteria. However, spectrum characteristics can alter due to the dynamic nature of PU traffic and network parameters. Thus, appropriate spectrum decisions and necessary interactions among application streams, need to be performed to meet the QoS provisioning. Several QoS supportive resource allocation schemes are reported in the literature. Most of the approaches assumed that PU traffic characteristics are known by SUs. However, this may not be true in general. Hence, PU traffic parameter estimation needs to be done often [24]. Such accurate estimation is necessary to guarantee QoS criteria in terms of disruption caused to the PUs. Overall, further investigations are needed to design QoS supportive RA schemes for CRN based on different criteria.

6.2.7 Cognitive Heterogeneous Networks (HetNets)

Cognitive heterogeneous networks (HetNets) are an attractive solutions for expanding network capacity to multiple spectrum access technologies, network structures and communication protocols. Some applications such as tactical applications in military communications [25], medical applications in CR-based wireless body area networks [26], are few interesting recent applications of cognitive HetNets. A thorough study of RA schemes with the capability of cognitive HetNets is still an open area of research. However, only a few works address the issues in a coexistent heterogeneous CRNs scenario [27]. Since the system characteristics of these cognitive HetNets, are different in several categories such as spectrum sensing aspects, PU detection ability, hardware capacity, fairness issue, all these categories need to be explored while designing MAC structure. A scenario of coexistent heterogeneous CRNs with collision-based PUs and fairness issue are discussed in [28]. Further research towards opportunistic 3G/4G/5G spectrum sharing involve precise evaluation of the scenario conditions, in terms of terminal confinement, link maintenance, and user preferences, and will also need to consider face side aspects such as security supervision and privacy preservation. The rising requirement of wireless data traffic and the scarceness of accessible radio spectrum, will extend 3GPP's LTE (Release 8) and LTE-Advanced (Release 10) to unlicensed bands (5 GHz). Thus dynamic spectrum management for LTE [29] and LTE-Advanced [30, 31] mobile communication network as a CR-ready technology, become challenging and provide an active area for research.

References

1. T. Yucek, H. Arslan, A survey of spectrum sensing algorithms for cognitive radio applications. IEEE Commun. Surv. Tutorials **11**, 116–130 (2009)
2. S. Hussain, X. Fernando, Closed-form analysis of relay-based cognitive radio networks over Nakagami-m fading channels. IEEE Trans. Veh. Technol. **63**, 1193–1203 (2014)
3. S. Hussain, X. Fernando, Performance analysis of relay-based cooperative spectrum sensing in cognitive radio networks over non-identical Nakagami-m channels. IEEE Trans. Commun. **62**, 2733–2746 (2014)
4. Y. Wang, P. Ren, Q. Du, Z. Su, Resource allocation and access strategy selection for QoS provisioning in cognitive networks, in *Proceedings of IEEE International Conference on Communications (ICC)* (2012), pp. 4637–4641
5. Y. Li, A. Nosratinia, Hybrid opportunistic scheduling in cognitive radio networks. IEEE Trans. Wirel. Commun. **11**, 328–337 (2012)
6. V. Srivastava, M. Motani, Cross-layer design: a survey and the road ahead. IEEE Commun. Mag. **43**, 112–119 (2005)
7. Y. Peng, F. Xiang, H. Long, The research of cross-layer architecture design and security for cognitive radio network, in *Proceeding of IEEE International Symposium on Information Engineering and Electronic Commerce (IEEC)* (2009), pp. 603–607
8. K. Ren, H. Zhu, Z. Han, R. Poovendran, Security in cognitive radio networks, in *Proceeding of IEEE Network*, vol. 27 (2013), pp. 2–3
9. R.K. Sharma, D.B. Rawat, Advances on security threats and countermeasures for cognitive radio networks: a survey. IEEE Commun. Surv. Tutorials **17**, 1023–1043 (2015)
10. A. He, K.K. Bae, T.R. Newman, J. Gaeddert, K. Kim, R. Menon, L. Morales-Tirado, J.J. Neel, Y. Zhao, J.H. Reed, W.H. Tranter, A survey of artificial intelligence for cognitive radios. IEEE Trans. Veh. Technol. **59**, 1578–1592 (2010)
11. I. Christian, S. Moh, I. Chung, J. Lee, Spectrum mobility in cognitive radio networks. IEEE Commun. Mag. **50**, 114–121 (2012)
12. X. Liu, Z. Ding, ESCAPE: a channel evacuation protocol for spectrum-agile networks, in *Proceeding of IEEE International Symposium New Frontiers in Dynamic Spectrum Access Networks (DySPAN)* (2007), pp. 292–302
13. L. Giupponi, and A.I. Pérez-Neira, Fuzzy-based spectrum handoff in cognitive radio networks, in *Proceeding of IEEE International Conference on Cognitive Radio Oriented Wireless Networks and Communications (CrownCom)* (2008), pp. 1–6
14. I. Butun, A.C. Talay, D.T. Altilar, M. Khalid, R. Sankar, Impact of mobility prediction on the performance of cognitive radio networks, in *Proceeding of IEEE Wireless Telecommunications Symposium (WTS)* (2010), pp. 1–5
15. X. Xing, T. Jing, W. Cheng, Y. Huo, X. Cheng, Spectrum prediction in cognitive radio networks. IEEE Wirel. Commun. **20**, 90–96 (2013)
16. I.A. Akbar, W.H. Tranter, Dynamic spectrum allocation in cognitive radio using hidden Markov models: Poisson distributed case, in *Proceeding of IEEE SoutheastCon* (2007), pp. 196–201
17. V.K. Tumuluru, P. Wang, D. Niyato, A neural network based spectrum prediction scheme for cognitive radio, in *Proceeding of ICC*, Cape Town (2010), pp. 1–5

18. X. Xing, T. Jing, Y. Huo, H. Li, X. Cheng, Channel quality prediction based on Bayesian inference in cognitive radio networks, in *Proceeding of IEEE INFOCOM*, Turin (2013), pp. 1465–1473
19. Z. Wen, T. Luo, W. Xiang, S. Majhi, Y. Ma, Autoregressive spectrum hole prediction model for cognitive radio systems, in *Proceeding of IEEE International Conference on Communications Workshops* (2008), pp. 154–157
20. S. Nejatian, S.K. Syed-Yusof, N.M.A. Latiff, V. Asadpour, Integrated handoff management in cognitive radio mobile ad hoc networks, in *Proceeding of IEEE International Symposium on Personal Indoor and Mobile Radio Communications (PIMRC)* (2013), pp. 2887–2892
21. D. Gözüpek and F. Alagöz, An opportunistic pervasive networking paradigm: multi-hop cognitive radio networks, in *Pervasive Computing and Networking*, ed. by M.S. Obaidat, M. Denko I. Woungang (Wiley, Chichester, 2011). https://doi.org/10.1002/9781119970422.ch7
22. K.M. Rabbi, D.B. Rawat, M.A. Ahad, T. Amin, Analysis of multi-hop opportunistic communications in cognitive radio network, in *Proceeding of IEEE SoutheastCon*, Fort Lauderdale (2015), pp. 1–8
23. M. Kartheek, V. Sharma, Providing QoS in a cognitive radio network, in *Proceeding of IEEE International Conference on Communication Systems and Networks (COMSNETS)* (2012), pp. 1–9
24. A. Sahoo, M. Souryal, Implementation of an opportunistic spectrum access system with disruption QoS provisioning and PU traffic parameter estimation, in *Proceeding of IEEE International Conference Wireless Communications and Networking Conference (WCNC)* (2015), pp. 1084–1089
25. M. Suojanen, J. Nurmi, Tactical applications of heterogeneous ad hoc networks – cognitive radios, wireless sensor networks and COTS in networked mobile operations, in *The Proceeding of International Conference on Advances in Cognitive Radio (COCORA)* (2014), pp. 1–5
26. R. Chávez-Santiago, I. Balasingham, Cognitive radio for medical wireless body area networks, in *In the Proceeding of IEEE International Workshop on Computer Aided Modeling and Design of Communication Links and Networks (CAMAD)* (2011), pp. 148–152
27. G.A. Shah, V.C. Gungor, O.K. Akan, A cross-layer design for QoS support in cognitive radio sensor networks for smart grid applications, in *Proceeding of IEEE International Conference on Communications (ICC)* (2012), pp. 1378–1382
28. Y.-C. Cheng, E.H. Wu, G.-H. Chen, A decentralized MAC protocol for unfairness problems in coexistent heterogeneous cognitive radio networks scenarios with collision-based primary users. IEEE Syst. J. **PP**, 1–12 (2015)
29. J. Naranjo, I. Viering, K. Friederichs, A cognitive radio based dynamic spectrum access scheme for LTE heterogeneous networks, in *Proceeding of IEEE Wireless Telecommunications Symposium (WTS)* (2012), pp. 1–7
30. J. Deaton, R. Irwin, L. DaSilva, The effects of a dynamic spectrum access overlay in LTE-advanced networks, in *Proceeding of IEEE Symposium on New Frontiers in Dynamic Spectrum Access Networks (DySPAN)* (2011), pp. 488–497
31. J.D. Naranjo, G. Bauch, A.B. Saleh, I. Viering, R. Halfmann, A dynamic spectrum access scheme for an LTE-advanced HetNet with carrier aggregation, in *Proceedings of International ITG Conference on Systems, Communication and Coding (SCC)* (2013), pp. 1–6

Appendix

A.1 Derivation of $\overline{P}_{f_i}$ for a Single-Relay System

Substituting $f_{G_i}(g)$ from (2.7) into (2.10) yields

$$\overline{P}_{f_i} = \left(\frac{m_i}{\overline{g}_i}\right)^{m_i} \frac{1}{\Gamma(m_i)} \int_0^\infty \frac{1}{\Gamma(m_i)} \Gamma\left(m_i, \frac{m_i\lambda}{N_0(1+A_ig)}\right)$$
$$g^{m_i-1} e^{-\left(\frac{m_i}{\overline{g}_i}\right)g} dg. \tag{A.1}$$

Using the series summation $\Gamma(m, x) = (m-1)! e^{-x} \sum_{n=0}^{m-1} \frac{x^n}{n!}$ [1, Eq. 8.352.2], with the fact that $(m-1)! = \Gamma(m)$, (A.1) becomes

$$\overline{P}_{f_i} = \left(\frac{m_i}{\overline{g}_i}\right)^{m_i} \frac{1}{\Gamma(m_i)} \sum_{q=0}^{m_i-1} \frac{1}{q!} \left(\frac{m_i\lambda}{N_0}\right)^q$$
$$\int_0^\infty (1+A_ig)^{-q} g^{m_i-1} e^{-\frac{m_i\lambda}{N_0(1+A_ig)}} e^{-\left(\frac{m_i}{\overline{g}_i}\right)g} dg. \tag{A.2}$$

Using $e^{-\frac{a}{b+x}} = \sum_{k=0}^\infty \frac{(-1)^k a^k}{k!(b+x)^k}$, we have

$$\overline{P}_{f_i} = \left(\frac{m_i}{\overline{g}_i}\right)^{m_i} \frac{1}{\Gamma(m_i)} \sum_{q=0}^{m_i-1} \frac{1}{q!} \left(\frac{m_i\lambda}{N_0}\right)^q$$
$$\sum_{n=0}^\infty \frac{(-1)^n}{n!} \left(\frac{m_i\lambda}{N_0}\right)^n \int_0^\infty (1+A_ig)^{-(q+n)} g^{m_i-1} e^{-\left(\frac{m_i}{\overline{g}_i}\right)g} dg. \tag{A.3}$$

X. Fernando et al., *Cooperative Spectrum Sensing and Resource Allocation Strategies in Cognitive Radio Networks*, SpringerBriefs in Electrical and Computer Engineering, https://doi.org/10.1007/978-3-319-73957-1

The above integral is evaluated with the aid of $\int_0^\infty e^{-px}x^{q-1}(1+ax)^{-v}dx = \frac{\Gamma(q)}{a^q}U(q;q+1-v;\frac{p}{a})$ [1, Eq. 3.383.5] to yield the result shown in (2.11).

A.2 Derivation of $\overline{P}_{fsc}$

Under non-identical fading, the general form for the PDF $f_{G_{max}}(g)$ is given by [2, 3]

$$f_{G_{max}}(g) = \sum_{j=1}^{L} f_{G_j}(g) \prod_{\substack{i=1 \\ i \neq j}}^{L} F_{G_i}(g) \tag{A.4}$$

where, $F(.)$ denotes the CDF. The above PDF is evaluated for integer values of m in [4] as follows

$$f_{G_{max}}(g) = \sum_{k=0}^{L-1} \frac{(-1)^k}{k!} \sum_{j=1}^{L} \frac{\beta_j^{m_j}}{\Gamma(m_j)} \underbrace{\sum_{n_1=1}^{L} \cdots \sum_{n_k=1}^{L}}_{n_1 \neq n_2 \neq \cdots n_k \neq j} \sum_{l_1=0}^{m_{n_1}-1}$$

$$\cdots \sum_{l_k=0}^{m_{n_k}-1} \left(\prod_{t=1}^{k} \frac{\beta_{n_t}^{l_t}}{l_t!} \right) g^{v_{kj}-1} e^{-\mu_{kj}g}. \tag{A.5}$$

Using (A.5), (2.14) can be expressed as follows

$$\overline{P}_{fsc} = \sum_{k=0}^{L-1} \frac{(-1)^k}{k!} \sum_{j=1}^{L} \frac{\beta_j^{m_j}}{\Gamma(m_j)} \underbrace{\sum_{n_1=1}^{L} \cdots \sum_{n_k=1}^{L}}_{n_1 \neq n_2 \neq \cdots n_k \neq j} \sum_{l_1=0}^{m_{n_1}-1} \cdots \sum_{l_k=0}^{m_{n_k}-1} \left(\prod_{t=1}^{k} \frac{\beta_{n_t}^{l_t}}{l_t!} \right)$$

$$\int_0^\infty \frac{1}{\Gamma(m_{SC})} \Gamma\left(m_{SC}, \frac{m_{SC}\lambda}{N_0(1+A_{SC}g)} \right) g^{v_{kj}-1} e^{-\mu_{kj}g} dg. \tag{A.6}$$

Using the series summation $\Gamma(m, x) = (m-1)! e^{-x} \sum_{n=0}^{m-1} \frac{x^n}{n!}$ and the fact that $e^{-\frac{a}{b+x}} = \sum_{k=0}^{\infty} \frac{(-1)^k a^k}{k!(b+x)^k}$ in a way similar to Appendix A.1, we have

$$\overline{P}_{fsc} = \sum_{k=0}^{L-1} \frac{(-1)^k}{k!} \sum_{j=1}^{L} \frac{\beta_j^{m_j}}{\Gamma(m_j)} \underbrace{\sum_{n_1=1}^{L} \cdots \sum_{n_k=1}^{L}}_{n_1 \neq n_2 \neq \cdots n_k \neq j} \sum_{l_1=0}^{m_{n_1}-1}$$

$$\cdots \sum_{l_k=0}^{m_{n_k}-1}\left(\prod_{t=1}^{k}\frac{\beta_{n_t}^{l_t}}{l_t!}\right)\sum_{q=0}^{m_j-1}\frac{1}{q!}\left(\frac{m_{SC}\lambda}{N_0}\right)^q$$

$$\sum_{n=0}^{\infty}\frac{(-1)^n}{n!}\left(\frac{m_{SC}\lambda}{N_0}\right)^n\int_0^{\infty}(1+A_{SC}g)^{-(q+n)}g^{\nu_{kj}-1}e^{-\mu_{kj}g}dg. \quad (A.7)$$

Using $\int_0^{\infty}e^{-px}x^{q-1}(1+ax)^{-\nu}dx = \frac{\Gamma(q)}{a^q}U(q; q+1-\nu; \frac{p}{a})$ [1, Eq. 3.383.5], to evaluate the integral in (A.7), the desired result is obtained as in (2.15).

A.3 Derivation of $\overline{P}_{f_{MRC}}$

Since g_i's are independently distributed and each follows the gamma distribution given in (2.7), the moment generating function of $\sum_i^L g_i$, is given by

$$M_G(t) = \prod_{i=1}^{L}M_{G_i}(t) = \prod_{i=1}^{L}(1-\beta_i t)^{-m_i} \quad (A.8)$$

where $M_{G_i}(t), i = 1, 2, \cdots, L$ is the moment generating function of the channel gain g_i. To derive a PDF for the random variable R, we define a new random variable $w_i = A_i g_i$. Obviously, $\beta_{w_i} = \frac{m_i}{A_i g_i} = \frac{\beta_i}{A_i}$ and the moment generating function of $R = \sum_i^L w_i$ can be expressed as

$$M_R(t) = \prod_{i=1}^{L}M_{W_i}(t) = \prod_{i=1}^{L}(1-\frac{\beta_i}{A_i}t)^{-m_i} \quad (A.9)$$

Using a general partial fraction technique to compute the inverse of $M_R(t)$ and following the same steps as in [5], the PDF of R for integer values of m_i is obtained as follows

$$f_R(r) = \left[\prod_{j=1}^{L}\left(-\frac{\beta_i}{A_i}\right)^{m_j}\right]\sum_{j=1}^{L}\sum_{v=1}^{m_j}\frac{(-1)^v b_{jv}r^{v-1}e^{-\frac{\beta_j}{A_j}r}}{(v-1)!} \quad (A.10)$$

The PDF given in (A.10) requires all the β_i's to be distinct and not necessary to be integers. If some of the β_i's are equal then the corresponding factors can be combined. Substituting (A.10) into (2.26) yields

$$
\overline{P}_{f_{MRC}} = \left[\prod_{j=1}^{L} \left(-\frac{\beta_i}{A_i} \right)^{m_j} \right] \sum_{j=1}^{L} \sum_{v=1}^{m_j} \frac{(-1)^v b_{jv}}{(v-1)!}
$$

$$
\int_0^\infty \frac{1}{\Gamma(m_j)} \Gamma\left(m_j, \frac{m_j \lambda}{N_0(1+r)} \right) r^{v-1} e^{-\frac{\beta_j}{A_j} r} \, dr. \tag{A.11}
$$

Following the same steps in Appendices A.1 and A.2, to evaluate the above integral, (A.11) becomes

$$
\overline{P}_{f_{MRC}} = \left[\prod_{j=1}^{L} \left(-\frac{\beta_i}{A_i} \right)^{m_j} \right] \sum_{j=1}^{L} \sum_{v=1}^{m_j} (-1)^v b_{jv}
$$

$$
\sum_{q=0}^{m_j-1} \frac{1}{q!} \left(\frac{m_i \lambda}{N_0} \right)^q \sum_{n=0}^{\infty} \frac{(-1)^n}{n!} \left(\frac{m_i \lambda}{N_0} \right)^n U(v; v+1-q-n; \frac{\beta_j}{A_j}). \tag{A.12}
$$

A.4 Derivation of $P_{m,n}^*$

In order to find the solution of the optimization problem depicted in (5.9), Lagrange multipliers $(\alpha, \beta, \gamma, \delta, \psi)$ are used corresponding to the five constraints. The Lagrange can be formed as

$$
L(\mathbf{X,T}, \alpha, \beta, \gamma, \delta, \psi) = \Delta B \sum_{m=1}^{M} \sum_{n=1}^{N} x_{m,n} \log \left(1 + \frac{t_{m,n}}{x_{m,n}} \Gamma_{m,n} \right) - \sum_{m=1}^{M} \sum_{n \in \mathcal{N}_m} \alpha_n \left(\sum_{m=1}^{M} \sum_{n=1}^{N} t_{m,n} - P_T \right)
$$

$$
- \sum_{m=1}^{M} \beta_m \left(\sum_{n=1}^{N} t_{m,n} - P_{m,max} \right) - \sum_{m=1}^{M} \sum_{n \in \mathcal{N}_m} \gamma_n \left(t_{m,n} - \bar{P}_{m,n} \right) - \sum_{k=1}^{K} \delta_k \left(\sum_{m=1}^{M} \sum_{n=1}^{N} t_{m,n} I_{m,n}^{(k)} - I_{th} \right)
$$

$$
- \sum_{m=1}^{M} \psi_m \left(\mathfrak{R}_{m,min} - \sum_{n=1}^{N} x_{m,n} \mathfrak{R}_{m,n} \right) \tag{A.13}
$$

The Lagrangian dual problem can be rewritten as follows:

$$
D(\alpha, \beta, \gamma, \delta, \psi) = \max_{\mathbf{X,T}} \sum_{m=1}^{M} \sum_{n=1}^{N} \left\{ (1 + \psi_m) x_{m,n} \log \left(1 + \frac{t_{m,n}}{x_{m,n}} \Gamma_{m,n} \right) - \left(\alpha_n + \beta_m + \gamma_n + \sum_{k=1}^{K} \delta_k \right) \right.
$$

$$
\left. I_{m,n}^{(k)} \right) t_{m,n} \right\} + \sum_{m=1}^{M} \sum_{n \in \mathcal{N}_m} \alpha_n P_T + \sum_{m=1}^{M} \beta_m P_{m,max} + \sum_{m=1}^{M} \sum_{n \in \mathcal{N}_m} \gamma_n \bar{P}_{m,n} + \sum_{k=1}^{K} \delta_k I_{th} - \sum_{m=1}^{M} \psi_m \mathfrak{R}_{m,min}
$$

$$= \max_{\mathbf{X}} \sum_{m=1}^{M} \sum_{n=1}^{N} \max_{\mathbf{P}} \left[x_{m,n} \left\{ (1+\psi_m) \log \left(1+P_{m,n} \Gamma_{m,n} \right) - \left(\alpha_n + \beta_m + \gamma_n + \sum_{k=1}^{K} \delta_m I_{m,n}^{(k)} \right) P_{m,n} \right\} \right]$$

$$+ \sum_{m=1}^{M} \sum_{n \in \mathcal{N}_m} \alpha_n P_T + \sum_{m=1}^{M} \beta_m P_{m,max} + \sum_{m=1}^{M} \sum_{n \in \mathcal{N}_m} \gamma_n \bar{P}_{m,n} + \sum_{k=1}^{K} \delta_k I_{th} - \sum_{m=1}^{M} \psi_m \Re_{m,min}$$

$$= \max_{\mathbf{X}} \sum_{m=1}^{M} \sum_{n=1}^{N} \max_{\mathbf{P}} \left\{ x_{m,n} \Psi(P_{m,n}) \right\} + \sum_{m=1}^{M} \sum_{n \in \mathcal{N}_m} \alpha_n P_T + \sum_{m=1}^{M} \beta_m P_{m,max} + \sum_{m=1}^{M} \sum_{n \in \mathcal{N}_m} \gamma_n \bar{P}_{m,n}$$

$$+ \sum_{k=1}^{K} \delta_k I_{th} - \sum_{m=1}^{M} \psi_m \Re_{m,min}$$

$$(A.14)$$

with $\Psi(P_{m,n})$ being defined as follows:

$$\Psi(P_{m,n}) = (1 + \psi_m) \log \left(1 + P_{m,n} \Gamma_{m,n} \right) - \left(\alpha_n + \beta_m + \gamma_n + \sum_{k=1}^{K} \delta_k I_{m,n}^{(k)} \right) P_{m,n}$$

$$(A.15)$$

To maximize (A.14) for any given $x_{m,n}$, (A.15) is differentiated with respect to $P_{m,n}$ and set the result to 0. This yields

$$P_{m,n}^* = \left[\frac{1 + \psi_m}{\ln \left(\alpha_n + \beta_m + \gamma_n + \sum_{k=1}^{K} \delta_k I_{m,n}^{(k)} \right)} - \frac{1}{\Gamma_{m,n}} \right]^+ \qquad (A.16)$$

References

1. I.S. Gradshteyn, I.M. Ryzhik, *Table of Integrals, Series and Products*, 6th edn. (Academic Press, London, 2000)
2. R. Kwan, C. Leung, General order selection combining for Nakagami and Weibull Fading Channels. IEEE Trans. Wirel. Commun. **6**(6), 2027–2033 (2007)
3. R.J. Vaughan, W.N. Venables, Permanent expressions for order statistics densities. J. R. Stat. Soc. **34**(2), 308–310 (1972)
4. R. Kwan, C. Leung, Selection diversity in non-identically distributed Nakagami Fading channels, in *Proceedings of Sarnoff Symposium (SARNOFF'05)*, Princeton (2005)
5. A.M. Mathai, Storage capacity of a dam with gamma type inputs. Ann. Inst. Stat. Math. **34**(3), 591–570 (1982)

Printed in the United States
By Bookmasters